园林深处是江南
Encountering Jiangnan
in the Gardens

丛书主编 章明
Chief Editor ZHANG Ming

# 行走江南园林 · 何园篇

杨晨 著

# Wandering Through Jiangnan Gardens: Ho Family Garden

YANG Chen

同济大学出版社
TONGJI UNIVERSITY PRESS
· 上海 ·

# 总序
## 江南园林的时空叙事与文化转译

### 缘起：园林作为文明对话的媒介

2024年春，法国凡尔赛，一场名为“行走江南园林”的展览以东方园林的曲径通幽，回应了西方几何园林的理性秩序。展览以“园林·景”的物质建构与“园林·境”的精神投射为双轴，构建起跨文化的阐释场域。展览中，豫园的“城市山林”、方塔园的“与古为新”、网师园的“壶中天地”，经由模型、影像与当代艺术的转译，成为西方观众解码中国园林美学的密钥。而这一文化事件的深层意义，恰在于其揭示了园林作为“可阅读的空间文本”与“可传播的文化基因”的双重属性——这也正是“行走江南园林”系列丛书编纂的学术起点。

### 六园考辨：江南园林的类型学切片

江南园林是中国古典园林的重要分支，以其精巧的布局、诗意的境界和深厚的文化内涵闻名于世。其营造注重“虽由人作，宛自天开”的美学原则，通过明旨相地、立意问名、布局借景、掇山置石、承续创新，将自然山水浓缩于方寸之间，形成“咫尺山林”的独特空间。江南园林不仅是文人雅士寄情山水的载体，更是中国传统哲学、艺术与生活方式的物质凝结，体现了“天人合一”的宇宙观和“诗画同源”的审美追求。江南园林并非均质化的美学符号，其地域特征、营造年代、主人身份及功能属性的差异，形成了独特的类型学谱系，“行走江南园林”丛书的第一辑选取网师园、寄畅园、何园、豫园、秋霞圃、方塔园，覆盖了江南私家园林中宅邸园林、山麓别墅、商贾园林等主要类型以及传承发展的现代园林，恰构成理解这一谱系的六把钥匙。

## 网师园

网师园以 0.6 公顷的微型尺度，完成了一场精妙的空间演绎。其“宅园一体”的布局中，轿厅与万卷堂构成礼仪轴线，而殿春簃则转向文人雅集功能。美国大都会博物馆的“明轩”移植，不仅复制了其物理空间，更引发了关于园林文化“可移植性”的学术论争。园中“竹外一枝轩”的命名源自苏轼诗句，揭示了文人园林作为“立体诗画”的本质属性。

## 寄畅园

园名取王羲之“取欢仁智乐，寄畅山水阴”诗意。作为现存唯一保持明代理水原貌的园林，寄畅园代表了山麓别墅园林的巅峰。此园最妙处，在“借”字：八音涧引惠山泉水，通过叠石控制水流声响；锦汇漪倒映锡山龙光塔，将三公里外的塔影纳入园景。康熙六巡江南，七次驻跸寄畅园，留下“山色溪光”御笔，揭示了其“真山真水”的营造智慧。这种“外师造化，中得心源”的造园理念，与计成《园冶》中“虽由人作，宛自天开”形成理论互文。

何园

何芷舠 1883 年建造的寄啸山庄，标志着扬州园林从盐商彰显财力转向中西合璧。1 500 米复道回廊形成立体交通，玉绣楼的法式百叶窗与中式雀替共处一室。其假山采用石涛“卷云皴”画意，而片石山房更是被考证为石涛叠山孤本。何园的空间叙事，暗合晚清“中体西用”的文化选择，是研究园林作为文化杂交载体的重要案例。

豫园

作为“海派园林”的肇始，豫园体现了明代商贾阶层对文人园林的模仿与重构。其空间格局曲折有致、虚实相生，龙墙蜿蜒分割四十余景，玉玲珑石“透漏皱瘦”，集湖石美学之大成。陈从周先生 1956 和 1986 年两次主持的修复工程，不仅恢复了潘允端“豫悦老亲”的初愿，更开创了“以园养园”的遗产活化模式。豫园的特别之处，在于其见证了上海从江南县城到国际都会的转型，是研究园林与城市互动的样本。

秋霞圃

嘉定秋霞圃由明代龚氏园、沈氏园、金氏园及宋代邑庙合并而成“三园一庙”的空间结构，分合有度，虚实相生，小中见大，再现天地，营造出城市山林的意境。桃花潭与凝霞阁构成南北对景，池北假山采用黄石叠砌，与苏州湖石假山形成对峙。园内保留的明代楠木厅，梁架比例严谨，装饰素雅，与清康熙年间增建的丛桂轩形成时代层积。其布局既遵循传统礼制，又兼具文人隐逸的灵动，规整性与文人园林的写意性交织，折射出礼制与逸趣的张力。

方塔园

冯纪忠1978年设计的方塔园，标志着中国园林从“私家宅园”向“公共空间”的范式转型。其以宋代方塔、明代照壁为历史锚点，通过现代混凝土框架重构传统空间序列。北门入口的堑道设计，暗合《园冶》中“欲扬先抑”的造园法则；何陋轩的竹构屋顶，则是对“茅茨不剪”的诗意转译。这种“新文言”式的设计语言，使方塔园成为研究园林现代性转化的典型案例。

这六座园林，在时间轴上串联起江南园林从明至今的演变轨迹，在空间轴上构建起从宅邸园林到山麓别墅，从文人园林到商贾复合的类型谱系，在营造技艺上则涵盖了相地、立基、掇山、理水、建筑布局、植物配置、借景等体系，堪称立体的《园冶》注解。园林的传播，本质是空间经验的转译。我们试图以六座园林为经纬，织就一张江南文化的认知之网：网师园的壶天密语，寄畅园的真山真水，何园的中西变奏，豫园的市井烟火，秋霞圃的婉约与端庄，方塔园的现代性实验——它们共同构成了中国园林的“类型学标本”。

## 书写范式：三重维度下的园林认知

丛书突破传统园林著述的单一视角，以“说园—游园—品园”的复合叙事结构，构建起理解园林的认知框架。“说园”叙述历史的考据与空间的分析；“游园”关注身体的丈量与视线的编织，体现了园林审美的体验性与身临其境；“品园”内观诗心的映照与哲学的顿悟：暗合童寯先生“园林三境论”所对应的“观”“品”“悟”的认知层次。同时融入当代数字人文的研究方法，辅以时间和空间两个“行走”的维度，带领读者穿梭于园林的文化历史和场景营造之间，不仅实现了园林文化的大众传播目的，又能满足以园林文化为主题的城市行走的读者诉求。

正如法国汉学家弗朗索瓦·朱利安所言：“园林是中国人处理‘之间’的艺术。”这套丛书试图揭示的，正是这种“之间性”——在传统与现代之间，在地方性与全球性之间，在物质空间与精神图景之间。

作为方法论的园林

本丛书汇聚风景园林、建筑史学、文化地理、艺术哲学等领域的学者，继承和发扬了江南园林从理论研究与修复保护，到文化实践，再到大众及国际传播的学术精神与路径，在“大众传播”与“学术深度”之间寻求平衡点。丛书以中英双语呈现，采用便携典雅的装帧设计，既是对展览的文献延伸，更是对江南园林文化谱系的一次系统性建构。其核心命题在于：如何将园林的“空间语法”转化为“文本语法”，如何让叠山理水的营造技艺升华为可被全球读者理解的文化叙事。

从计成《园冶》的“虽由人作，宛自天开”，到冯纪忠“与古为新”的现代诠释，江南园林始终是中国人处理时空关系的哲学实验室。“始于出版，不止于出版”——“行走江南园林”系列丛书的目标，不在于记录六座园林的物质形态，而在于实现以园林文化为载体的中华传统文化的科普教育、大众传播和国际传播，更在于揭示其背后“文化的可生长性”——这种生长性，体现在全球读者通过此书展开的文化想象。

章明

2025 年春于上海南京西路

# FOREWORD

## Jiangnan Gardens: Spatial-Temporal Narratives and Cultural Semiotic Translation

### Motivation: Gardens as a Medium for the Dialogue among Civilizations

In mid-spring 2024, an exhibition titled "Wandering Through Jiangnan Gardens" was held in Versailles, France, which showcased the winding paths and secluded vistas characteristic of Eastern gardens, providing a lyrical contrast to the structured order of Western geometric landscapes. Framed around two main themes— "Garden as Scenery," which reflects the psychological aspects of garden construction, and "Garden as Realm," which evokes a sense of spiritual projection—the exhibition fostered engaging cross-cultural dialogue. At the exhibition, a collection of images representing Chinese garden landscapes was semiotically translated into architectural models, immersive video art, and contemporary artworks. These included the "urban forest" motif from Yu Garden, the "renewal with the ancient" theme from the Shanghai Fangta Garden, and the "Heaven and Earth in a Gourd" concept seen in the Master-of-Nets Garden. These various elements serve as a key to understanding the aesthetics of Chinese gardens for Western audiences. They demonstrate that gardens and their landscapes can function as both a readable spatial text and a transmissible cultural essence.

This insight inspired the editorial vision and academic motivation for the book series *Wandering Through Jiangnan Gardens*.

## Six Gardens in Comparison: A Typological Study of Jiangnan Gardens

The Jiangnan garden represents a unique branch of classical Chinese landscape art, renowned for their refined spatial rhythms, poetic atmosphere, and profound cultural significance. Built on the aesthetic principle that "though artificially crafted, it appears divinely natural," its creation involves a thoughtful orchestration of site selection, thematic conception and poetic naming, spatial choreography and borrowed scenery, the artful placement of rocks and water, and a continuous dialogue between tradition and innovation. Through these techniques, the majesty of nature is distilled into intimate, human-scaled settings, embodying a distinctly Jiangnan vision of "mountains and forests that feel within reach."

More than refined retreats for scholars and literati, Jiangnan gardens have materialized the philosophical worldview, artistic sensibilities, and lifestyle of traditional China. They reflect a cosmology rooted in the "harmony between humanity and nature," and an aesthetic ideal based on the "shared origins of poetry and painting." Rather than presenting a monolithic aesthetic, they exhibit a rich typological diversity shaped by variations in geography, historical period, patron identity, and intended function.

The first collection of the series highlights six representative sites, namely, the Master-of-Nets Garden, the Jichang Garden, the Ho Family Garden, the Yu Garden, the Qiuxia Garden, and the Shanghai Fangta Garden, which encompass the principal typologies of Jiangnan private gardens—residential estates, hillside villas, mercantile gardens—alongside their contemporary successors, offering six distinct keys to decipher this garden genealogy.

## Master-of-Nets Garden, Suzhou

Though occupying just 0.6 hectares, the Master-of-Nets Garden achieves remarkable spatial complexity through masterful design. In its distinctive "residence-and-garden-in-one" layout, the Sedan-Chair Hall (*Jiao Ting*) and Hall of Ten-Thousand Volumes (*Wanjuan Tang*) form a ceremonial axis, while the Late Spring Retreat (*Dianchun Yi*) functions as a refined space for literati gatherings. Notably, it was replicated and reinterpreted as the Ming Hall (Astor Court) at the Metropolitan Museum of Art in New York, sparking scholarly discussions on the cultural transplantability of Chinese gardens. The name of the veranda, "a Twig Beyond Bamboos," derived from a poem by Su Shi (1037–1101), highlighted the essence of the literati garden as a form of "three-dimensional poetry and painting."

## Jichang Garden, Wuxi

The Jichang Garden, literally the Delight-Conveying Garden, echoes a verse composed by Wang Xizhi (303–361), the calligraphy sage in the 3rd century: "In the embrace of benevolence and wisdom, joy is found; amidst the tranquil shade of hills and streams, the mind is unbound." As the only surviving garden preserving an original Ming-era water feature, it represents the pinnacle of hillside villa landscape. Its brilliance lies in the concept of "borrowed scenery" (*jiejing*). On one hand, the Eight-note Stream channels spring water from Huishan Mountain, with rock formations shaping both the flow and sound of the water. On the other hand, the Brocade Ripple Pond (*Jinhui Yi*) mirrors the Longguang Pagoda atop Xishan Mountain in its shimmering water, seamlessly incorporating this distant landmark into the garden's overall composition. Historically, Emperor Kangxi (1654–1722), during his six southern inspection tours, resided seven times at the garden, where he inscribed "Mountain Hues and Stream Light" (*shanse xiguang*)—a testament to the garden's mastery in reproducing "genuine mountains and waters." The garden-making philosophy of "learning from natural creation while internalizing artistic conception" aligns closely with the principle proposed by Ji Cheng (ca. 1582–ca. 1642), a garden architect of the late Ming dynasty, in his treatise *The Craft of Gardens* (1634): "Though artificially crafted, it appears divinely natural." Such a theoretical intertextuality reveals the essence of Chinese landscape aesthetics.

## Ho Family Garden, Yangzhou

The Ho Family Garden, originally named the Whistling Scholars' Mountain Retreat (*Jixiao Shanzhuang*), was built by He Weijian in 1883. This garden represented a major change in garden construction in Yangzhou, shifting from simple displays of wealth by salt merchants to a harmonious blend of Chinese and Western influences. A notable feature of the garden is a 1,500-meter double-decker winding walkway system, which creates an intricate network of walkways. The Yuxiu Building prominently displays French-style shutters alongside Chinese decorative elements known as "bracket-connecting arms" (*queti*). The rockeries within the garden exhibit the "rolling-cloud texture strokes" (*juanyuncun*), a rock formation technique developed by Shi Tao (1642–1707), a legendary monk-painter, calligrapher, and art theorist of the early Qing dynasty. Among these, the rockery at the Small-Rock Mountain Retreat (*Pianshi Shanfang*) is particularly valued as a rare surviving example of his artistry in rockwork. The spatial design of the garden echoes the philosophy of "Western Means for Chinese Ends," which gained popularity during the late Qing dynasty (1644–1911). This makes the Ho Family Garden an important case study of cultural hybridity in garden design.

## Yu Garden, Shanghai

Yu Garden, the quintessential example of Shanghai-style gardens, showcases the merchants' imitation and reinterpretation of literati gardens from the Ming dynasty (1368–1644). Its intricate layout expertly balances solid structures with open spaces, featuring a dragon wall that meanders through the garden and divides the area into 40 scenic spots. The Exquisite Jade Rockery (*Yulinglong Shi*) is particularly remarkable, highlighting the typical characteristics of Lake Tai rocks, which are known for being perforated, porous, wrinkled, and slender. In 1956 and 1986, restoration projects led by Chen Congzhou (1918–2000), a renowned master of Chinese garden studies, not only revived the original vision of the founder, Pan Yunduan (1526–1601), who aimed to "bring joy to his parents," but also introduced the "garden-sustaining-garden" approach to heritage conservation. Yu Garden stands as a testament to Shanghai's transformation from a small Jiangnan town into an international metropolis, making it an important case study in the relationship between gardens and urban development.

## Qiuxia Garden, Shanghai

Located in the Jiading district, the Qiuxia Garden was created by merging three private gardens belonging to the Gong, Shen, and Jin families from the Ming dynasty, along with a county temple from the Song dynasty (960–1276). This integration resulted in a spatial structure known as "three gardens and one temple," characterized by a balanced interplay of separation and unity, solid and void, and a sense of vastness within a confined space. The design reflects the harmony between heaven and earth, capturing the poetic essence of an urban retreat. The Peach Blossom Pond (*Taohua Tan*) and the Rosy-Cloud-Gazing Belvedere (*Ningxia Ge*) create a north-to-south opposite view. The Yellow Mountain stones in the northern part of the pond contrast with the Lake Tai rockeries in the south, resulting in a striking juxtaposition. The garden also features a nanmu (*Phoebe zhennan*) hall from the Ming dynasty, notable for its precisely proportioned timber framework and understated ornamentation. This hall coexists with the Veranda of a Hillock Clustered with Osmanthus (*Conggui Xuan*), an addition from the Kangxi reign (1662–1722) of the Qing dynasty, showcasing layers of historical evolution. Overall, the layout adheres to traditional ceremonial norms while embracing the fluidity of the literati's contemplative spirit. It combines formality with the expressive aesthetics of literati gardens, embodying a dynamic tension between ritual propriety and poetic leisure.

## Shanghai Fangta Garden, Shanghai

Designed by Feng Jizhong (1915–2009) in 1978, the Shanghai Fangta Garden represents a significant transition in Chinese gardens from private estates to public spaces. Central to the park is the Square Pagoda (*Fangta*), which dates to the Song dynasty, alongside the Screen Wall from the Ming dynasty. These elements recreate traditional spatial sequences within a modern concrete framework. The sunken pathway at the north entrance subtly follows the principle of "restraining before revealing," as outlined in *The Craft of Gardens*. Additionally, the bamboo-structured roof of Helou Pavillion provides a poetic reinterpretation of the classical ideal of "leaving the thatched eaves untrimmed."

These design choices, infused with the spirit of "neo-classical Chinese expression," make the Shanghai Fangta Garden an intriguing case study in the contemporary evolution of Chinese garden aesthetics.

These six gardens illustrate the evolution of Jiangnan gardens from the Ming dynasty to now, creating a diverse range that includes residential gardens, hillside retreats, literati retreats, and merchant compounds. This variety represents a comprehensive system of spatial design, encompassing elements such as site selection, foundation work, rock formations, water features, architectural layouts, planting arrangements, and borrowed scenery. It can be viewed as a three-dimensional interpretation of *The Craft of Gardens*. The essence of garden culture is a semiotic translation of spatial experience. Our goal is to use these six gardens as a framework to weave a cognitive tapestry of Jiangnan culture: the tranquil cosmos of the Master-of-Nets Garden, the authentic mountains and living waters of the Jichang Garden, the Sino-Western elements of the Ho Family Garden, the urban vibrancy of Yu Garden, the elegance of the Qiuxia Garden, and the modernist approach of the Shanghai Fangta Garden. Together, they create a typological atlas of the Chinese garden tradition.

## Writing the Garden: Cognitive Frameworks Across Three Dimensions

This series adopts a tripartite narrative theme of "stories, routes, and appreciation" to establish a structured framework for understanding classical Chinese gardens. This approach resonates with the concept of the "three realms of garden perception"—visiting, appreciating, and understanding—proposed by Tung Chuin (1900–1983), a renowned landscape specialist. In this framework, the "stories" explore historical context and spatial analysis; the "routes" emphasize physical movement and visual experiences, highlighting the immersive aesthetics of the gardens; and the "appreciation" delve into poetic sensibility and philosophical insights. Furthermore, digital humanities methods invite readers to investigate the relationship between historical context and crafted scenes, encourage urban walks, and enhance mass communication centered on landscape heritage.

As the French sinologist François Jullien (1951– ) once observed, "The Chinese garden is the art of shaping the *l'entre* (literally, in-between)." This series aims to illuminate the interstitiality between tradition and modernity, the local and the global, the material and the spiritual.

## The Garden as Methodology

This series has brought together scholars from various fields, including landscape architecture, architectural history, cultural geography, and the philosophy of art. It continues the scholarly legacy of Jiangnan gardens by exploring theoretical concepts, conservation practices, cultural engagement, and international dialogue, aiming at a thoughtful balance between public accessibility and academic depth. Presented in both Chinese and English, the series has been designed in a compact yet elegant format. It serves as an archival extension of the exhibition held in Versailles while systematically exploring the artistic tradition of Jiangnan gardens. Additionally, it addresses critical questions, such as how to translate a "spatial grammar" into a "textual grammar" and how to convey the techniques of rockwork and water features as a cultural narrative that is intelligible to readers worldwide.

Jiangnan gardens have long served as a philosophical laboratory for exploring the concepts of space and time within Chinese culture, as illustrated in Ji Cheng's dictum—"Though artificially crafted, it appears divinely natural," and in Feng Jizhong's modern interpretation of "renewing the ancient." This book series goes beyond simply documenting the physical characteristics of six gardens; it acts as a tool for cultural education, public engagement, and international dialogue centered on garden culture, acting as a gateway to a broader understanding of Chinese tradition. At its core, the series showcases the generative potential of culture—a vitality that inspires the cultural imagination of global audiences.

ZHANG Ming
Spring 2025
West Nanjing Road, Shanghai

# 目　录

# CONTENTS

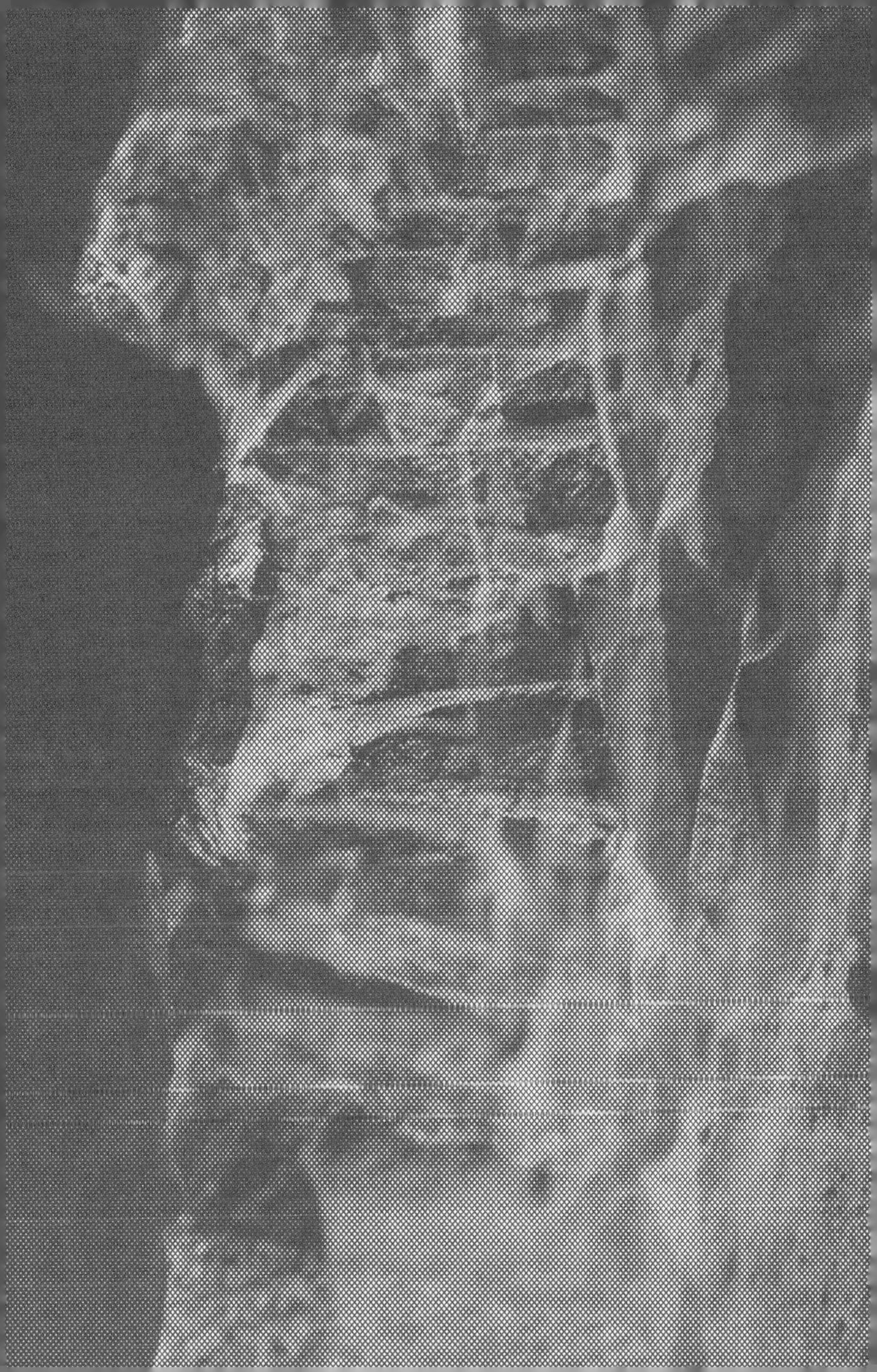

Amidst the myriads of classical Chinese gardens, one stands apart with a unique distinction: The Ho Family Garden in Yangzhou.

It features overlapping multi-level walkways, wall-backed rockeries, and a harmonious blend of the grandeur of Northern China with the elegance of Southern China, infused with a touch of Western charm.

From this historic and imposing mansion emerged generations of prominent figures, including pioneers of the Self-Strengthening Movement, the founding figure of China's Common-Law school, and academicians of the Chinese Academy of Sciences.

Hailed as "the Best Garden of the Late Qing Dynasty," this masterpiece of landscape artistry endures as a rare jewel of Chinese gardens.

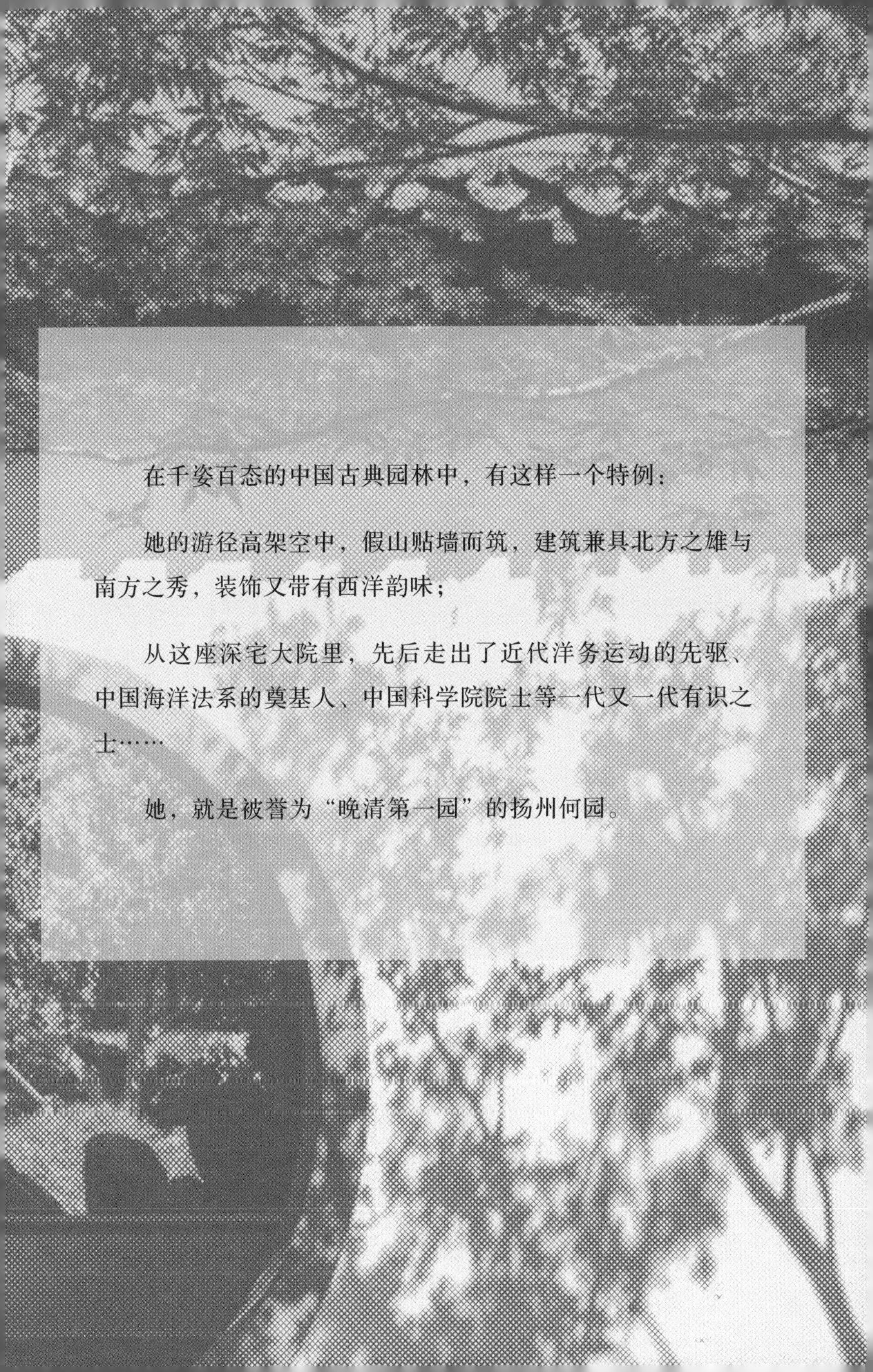

在千姿百态的中国古典园林中，有这样一个特例：

她的游径高架空中，假山贴墙而筑，建筑兼具北方之雄与南方之秀，装饰又带有西洋韵味；

从这座深宅大院里，先后走出了近代洋务运动的先驱、中国海洋法系的奠基人、中国科学院院士等一代又一代有识之士……

她，就是被誉为“晚清第一园”的扬州何园。

东阁梅花，扬州风月。
南塘野草，何氏山林。

Plum blossoms by the Eastern Belvedere,
echoing Yangzhou's scenic charm;
Wild grasses by the Southern Pond, steeped
in the Ho family's rustic retreat.

# Ho Family Garden

# 何园

**Location**
No. 66 Xuningmen Street, Yangzhou, China

**Year**
Started in 1883

**Site Area**
Approx. 1.15 hectares

**Area**
Approx. 6,000 square meters

**Protection Level**
Major Historical and Cultural Site Protected at the National Key Level

**位置**
江苏省扬州市
徐凝门大街 66 号

**年代**
1883 年始建

**占地面积**
约 1.15 公顷

**总建筑面积**
约 6 000 平方米

**级别**
全国重点文物保护单位

古运河
Yangzhou Ancient Canal
便益门
Bianyi Gate
广储门
Guangchu Gate
镇淮门
Zhenhuai Gate
拱辰门
Gongchen Gate
个园
Geyuan Garden
逸圃
Yipu Garden
通泗门
Tongsi Gate
利津门
Lijin Gate
珍园
Zhenyuan Garden
文昌中路
Middle Wenchang Road
小秦淮河
Little Qinhuai River
皮市街
Pishi Street
蔚圃
Weipu Garden
通济门
Tongji Gate
小盘谷
Xiaopangu Garden
徐凝门大街
Xuningmen Street
何園
Ho Family Garden
安江门
Anjiang Gate
挹江门
Yijiang Gate
徐凝门
Xuning Gate
明扬州城范围
The boundaries of Yangzhou City during the Ming Dynasty
N

# 说园

STORIES

## 走进晚清第一园

Into the Best Garden of the Late Qing Dynasty

Yangzhou, strategically situated at the confluence of the Yangtze and Huai River regions, is blessed with favorable natural conditions, abundant resources, and a rich cultural heritage. For centuries, it has been a prosperous hub, inspiring the saying: "With a hundred thousand strings of coins around my waist, I ride a crane to Yangzhou in haste." Liu Daguan (1753–1834), a literatus from the Qing dynasty, remarked that "Hangzhou excels in lakes and mountains, Suzhou in bustling markets, and Yangzhou in exquisite gardens and pavilions—these three cities stand as equals, none surpassing the others." Yangzhou is particularly renowned for its beautiful gardens, which reflect its historical rise and fall.

The garden construction in Yangzhou flourished during the Kangxi-Qianlong era (1662–1795), reaching its golden age. At that time, the city served as a vital hub for salt production, trade, transportation, and administration on both the northern and southern banks of the Huai River. Attracted by the promise of prosperity, merchants from across the empire flocked to Yangzhou, bringing immense wealth and a taste for refined living. These salt merchants built many private gardens, which were initially concentrated along Dongguan Street and later expanded to areas around Huayuan Lane and the north bank of the Yangzhou Ancient Canal.

During the Ming and Qing dynasties, this residential quarter lay along the Yangzhou Ancient Canal, specifically to the southeast of the New City (*xincheng*). The location was exceptionally advantageous with convenient access to both land and water routes and its proximity to the Salt Transport Office and bustling commercial districts. At the time, over a hundred private gardens dotted the city and its outskirts,

扬州地处江淮要冲，自然条件优越，物质资源丰沛，滋养出璀璨的文化盛景，乃千年繁华之地，自古有“腰缠十万贯，骑鹤上扬州”之说。清代文人刘大观称：“杭州以湖山胜，苏州以市肆胜，扬州以园亭胜，三者鼎峙，不可轩轾。”扬州以园林著称于世，而园林也是这座城市兴衰的写照。

扬州的园林建设在清代康乾时期达到鼎盛，此时的扬州是淮南、淮北盐业生产、贸易、运输和管理的中心，也是天下人向往的淘金热土，大批盐商聚集扬州，建造了大量园林。盐商的宅第园林多集中在东关街一带，后逐渐发展到花园巷、南河下一带。明清时期，这片居住区位于新城区东南部的古运河沿岸，水陆交通便利，靠近盐运署和商业地区，区位极佳。该时期城内城外有园林上百座，大多是盐商在城中用以私家游宴和移情养性的场所。清代嘉庆以后，由于海运发展和盐法改革等原因，扬州失去了作为内河航运重要港埠的核心地位，城市经济也逐渐衰落，盐商实力大不如前；咸丰年间，扬州城又因战事遭重大破坏。直到“同光中兴”时期，扬州盐业经济略有复苏，中国封建社会的回光返照也为扬州带来了私家园林建设的最后一轮高潮，出现了壶园、小盘谷、裕园、约园等园林，但规模已远不如前。

扬州何园，原名寄啸山庄，建于光绪九年（1883年），是中国封建社会晚期扬州大规模私家园林建造热潮中的唯一一

most built by salt merchants as venues for leisure, entertainment, and refined retreats.

After the reign of Emperor Jiaqing (1796–1820), Yangzhou's significance as a major inland port diminished due to the rise of maritime trade and reforms in the salt monopoly. The city's economy gradually declined, leading to a decrease in the power of salt merchants. During the Xianfeng reign (1851–1861), Yangzhou suffered extensive destruction due to warfare. It wasn't until the Tongzhi-Guangxu Restoration (1862–1894) that the salt economy began to see a modest revival. This period, marking the twilight of China's feudal society, also witnessed a final flourish in private garden construction, giving rise to gardens such as Fanghu Garden (*Hu Yuan*), Lesser Winding Gully (*Xiao Pangu*), Yu Garden, and Yue Garden—though none could match the grandeur of earlier times.

The Ho Family Garden in Yangzhou, originally named the Whistling Scholar's Mountain Retreat (*Jixiao Shanzhuang*), was built in 1883. It is significant for being the only grand garden constructed during the final period of large-scale private garden development in late imperial Yangzhou. As one of the last masterpieces of classical Chinese garden art, this garden exemplifies the refinement of a bygone era. In his poem, Chen Congzhou captured its ethereal charm and emphasizes its unique status among the classical gardens of Jiangnan, "Amidst the splendor of the land, Jiangnan gardens reign supreme, and Yangzhou claims two-thirds of the Bright Moonlight. Verdant ripples shimmer around the Water-Heart Pavilion, and one ascends the linking gallery to behold the vista."

座大型园林作品。园林大家陈从周先生曾赋诗盛赞："江南园林甲天下，二分明月在扬州。水心亭上春波绿，览胜来登一串楼"，既刻画了何园的灵秀风光，又彰显了其在江南园林中的独特地位。

何园整体鸟瞰

Aerial view of Ho Family Garden

# Site of the Twin Sophora Garden

The Ho Family Garden was not constructed all at once during the Guangxu reign (1875–1908); rather, it was gradually rebuilt on the foundation of older residential gardens. Historical records indicate that the original garden was the Twin Sophora Garden (*Shuanghuai Yuan*), built by Wu Jialong around 1703. This garden served as a retreat for the Wu family, located in the Nanhexia district, southeast of Yangzhou New Town. It extended from Huayuan Lane in the south to Diaojia Lane in the north.

Wu Jialong, a salt merchant from Anhui, was well known for his philanthropic efforts. During a widespread famine in 1738, he donated 7,000 taels of gold to help the victims, and four years later, he contributed an additional 3,000 taels. Wu also funded the restoration of important temples in Yangzhou, including the Precious Wheel Temple (*Baolun Si*) and the Serene Wisdom Temple (*Jinghui Si*). Historical documents indicate that during the Kangxi-Qianlong era, the Twin Sophora Garden boasted renowned architectural and scenic features, including the Rain-Listening Veranda (*Tingyu Xuan)*, the Studio of Bottles and Goblets (*Pinglei Zhai*), the Butterfly Hall (*Hudie Ting*), the Plum Hall (*Mei Lou*), the Waterside Gazebo (*Shui Xie*), and the Mountain Delight Hall (*Leshan Tang*).

The Twin Sphora Garden, also known as the Small-Rock Mountain Retreat (*Pianshi Shanfang*), derives its name from a

# 双槐园址

何园并非清代光绪年间一次建成，而是在前人宅园的基础上逐步改建而成的。历史记载，何园坐落地最早的园林营造是康熙中期（约1703年前后）吴家龙所创立的双槐园，它位于扬州新城东南的南河下片区，其范围南至花园巷，北至刁家巷，因园中两棵古槐树而得名，最初作为吴氏别业使用。吴家龙是定居在扬州的安徽盐商，以乐善好施而闻名。相传乾隆三年（1738年）恰逢荒年，吴家龙捐献了七千两黄金救济灾民，四年后（1742年）又捐三千余两。他还曾出资赞助维修扬州宝轮寺、静慧园等重要寺院。据文献记载，康乾时期的双槐园内有听雨轩、瓶櫑斋、蝴蝶厅、梅楼、水榭以及乐山堂等建筑，园林景观颇具盛名。

双槐园又名片石山房，取自园中一座别具特色的太湖石假山。清嘉庆十六年（1811年）《江都县续志》记载："片石山房在花园巷，吴家龙所辟。中有池，屈曲流。前为水榭，湖石三面环列，其最高者特立耸秀，一罗汉松踞其巅，几盈抱矣。"此处记载的湖石假山及罗汉松目前仍存于园中，罗汉松树龄已超340年，据推测是双槐园遗物。清代道光年间《履园丛话》记载："二厅之后，湫以方池。池上有太湖山子一座，高五六丈，甚为奇峭，相传为石涛和尚手笔。"今天何园里

distinctive rockery made of Lake Tai limestones. *The Sequel to the Gazetteer of Jiangdu County* (1811) notes that "the Small-Rock Mountain Retreat was established by Wu Jialong in Huayuan Lane. It features a winding pond within its grounds, and in front of this pond stands a waterside gazebo, surrounded on three sides by Lake Tai rockeries. The tallest rockery rises sharply and elegantly, and atop it grows a Buddhist pine tree whose trunk has a girth of one armspan." Both the rockery and the Buddhist pine are still present in the Ho Family Garden today. The pine tree, now over 340 years old, is believed to be a remnant of the original garden. According to *A Collection of Accounts at the Lü Garden*, completed during the Daoguang reign (1821–1850), it is mentioned that "behind the two buildings lies a square pond, and above it stands a Lake Tai rockery, 16 to 20 meters tall, jagged and imposing. It is said to have been crafted by the painter-monk Shi Tao." The current Small-Rock Mountain Retreat has been restored on the foundation of the Wu family's original estate, making it one of the most significant relics of Yangzhou gardens.

During the Qianlong reign (1736–1795), the Twin Sophora Garden was passed down to Wu Zhifu, the grandson of Wu Jialong. By the time of the Jiaqing reign (1796–1820), the garden had gradually fallen into disrepair, and much of its original beauty had diminished. Only a few structures, such as the Rain-Listening Veranda and the Waterside Gazebo, remained intact, and the garden was renamed the Twin Sophora Tea House (*Shuanghuai Chayuan*). Over the years, during the

的园中园“片石山房”，即是在吴氏园林的基础上，重新修缮整理而成，是目前最具历史价值的扬州园林遗存之一。

清代乾隆年间，双槐园传至吴家龙的孙子吴之黼，至嘉庆年间已逐渐荒废。这一时期园景收缩，只存在听雨轩、水榭等建筑，园名也改为“双槐茶园”。至道光年间，园林易主，曾一度成面馆、戏院，困于俗事，意境全无。同治至光绪年间，广东商人吴辉谟曾经在此短暂葺居，对园林及建筑做了一定的整理，恢复了其住宅花园的整体面貌。

片石山房小院入口
Entrance to the Small Garden of the
Small-Rock Mountain Retreat

Daoguang reign (1821–1850), the garden changed ownership and was transformed into a noodle shop and theater, resulting in a significant loss of its original elegance and atmosphere. From the Tongzhi to the Guangxu reigns (1862–1908), the garden was briefly owned by Wu Huimo, a merchant from Guangdong, who resided there for over a decade. He made efforts to restore the garden, partially bringing back its design as a cohesive residence and garden. However, in 1883, Wu sold the garden to He Weijian, a retired official, which marked the beginning of its transformation into the Ho Family estate.

古槐树根改造成的棋盘，犹存“双槐园”旧迹
A chessboard crafted from an ancient sophora root, still bearing traces of the Twin Sophora Garden

片石山房假山上的罗汉松
Buddhist Pine in front of the Small-Rock Mountain Retreat

# Ho Family Garden

He Weijian (1835–1908), with the courtesy name Ruchi and literary name Zhidao, held several important positions throughout his career, including the director of the Fujian and Yunnan Bureaus in the Ministry of Revenue, the Salt Commissioner in Wuchang, the Grain Intendant of Hubei, and the acting Provincial Judge of Hubei. In 1876, he was the circuit intendant for the Han-Huang-De area in Hubei Province, where he also oversaw the Hankou Customs and managed foreign trade affairs. Recognized for his integrity, he defended national sovereignty against foreign pressures while promoting economic growth and public welfare, earning widespread acclaim for his efforts. Upon acquiring the estate, he transformed it into a sprawling private garden, renaming it the Whistling Scholar's Mountain Retreat. The name was inspired by Tao Yuanming's poem "Back Home Again Chant," which includes the lines: "At the window, my presumptions drift away south... Maybe I'll climb East Ridge and chant, settling into my breath." This reflected his aspiration for a reclusive life, free from worldly concerns.

Locally known as "the Ho Family Garden," this site stands out as the largest and most prominent example of private garden structure during the final wave of classical garden construction in Yangzhou. Instead of demolishing the Twin Sophora Garden, He Weijian preserved key structures and landscapes such as the

# 何氏家宅

清代光绪九年（1883 年），吴辉谟在双槐园居住了十几年之后，将园林出售给放官回乡的官员何维键（1835—1908），由此开启了何园作为何家住宅园林的重要时期。何维键，字汝持，号芷舠，生于道光十五年（1835 年），先后任户部福建司、云南司郎中，任湖北补用道，两署湖北武昌盐法道，任湖北督粮道，署湖北按察使等。光绪二年（1876 年），何维键任湖北汉黄德道、江汉关监督，兼办通商事宜。在此期间他不畏强权、维护国家主权，关心民生、发展经济，获得了广泛赞誉。何氏将双槐园扩建成了规模庞大的私家宅园，并题名“寄啸山庄”，取陶渊明《归去来兮辞》中“倚南窗以寄傲”“登东皋以舒啸”句意，寓意园主的归隐之心。

何园乃“寄啸山庄”民间俗称，指何氏家族居住的住宅及花园，它是扬州古代私家园林最后一轮建设高潮中规模最大的作品。何氏在购入双槐园旧址之后，并未将其全部推倒，而是分为保留和新建两个步骤进行修整。据考证，何氏保留了部分建筑和景观，包括片石山房及楠木厅、骑马楼、东一楼、东二楼、东三楼等，也包括园林北部大花园的空间轮廓以及蝴蝶厅。而后，他拆除了双槐园中南部的建筑，新建了赏月楼、玉绣楼、煦春堂等，重建了读书楼、蝴蝶厅、北门等，并以

Small-Rock Mountain Retreat, the Nanmu Hall, the Horseback Riding Building (*Qima Lou*), and the three towers of the East Wing, along with the overall spatial contours of the Grand Garden and the Butterfly Hall to the north.

However, he did demolish some areas in the central south, where he built the Moon-Viewing Building, the Yuxiu Building, and Warm Spring Hall (*Xuchun Tang*). He also reconstructed the Study Studio, the Butterfly Hall, and the north gate, while renovating the landscape of the northern Grand Garden with a focus on the themes of seclusion and retreat. Additionally, He designed and constructed a multi-level walkway to connect buildings and courtyards, as well as the Ho Family Ancestral Hall in the southeastern corner. These enhancements contributed to the garden's cohesive and distinctive appearance. Once completed, the garden received widespread acclaim for its ingenious design, exquisite architecture, and elegant landscaping, being celebrated as "the finest garden in Yangzhou since the reigns of Xianfeng and Tongzhi (1851–1874)."

Although He Weijian and his family had lived in the garden for 18 years, they eventually relocated to Shanghai, where they invested in industries such as the Imperial Bank of China and the Hanyeping Coal and Iron Company. He passed away in 1908 at the age of 73. In October 1935, His eldest son, He Shenghao, moved back to the garden with his family due to financial constraints. However, in 1937, with the outbreak of Japanese full-scale aggression in China, the family—more than 20 people

何维键全身像，约翰·汤姆森摄于 1869 年，
现藏于伦敦维尔康姆收藏馆
He Weijian Portrait, Photo by John Thomson, 1869
Wellcome Collection, London, UK

across three generations—had to leave the garden again and return to Shanghai. During this time, the renowned garden specialist Tung Chuin visited the garden and later documented his impressions in his *Gazetteer of Jiangnan Gardens*: "The Ho Family Garden stands as the largest surviving private garden in Yangzhou. The residential quarters are located to the east, while the great hall lies at the center, surrounded by a series of courtyards. To the west, there are three rows of two-story, Western-style lodgings, and beyond these lie the main part of the garden, known as the Whistling Scholar's Mountain Retreat, which was built by the salt official He Weijian, but it has fallen into decline over the past 50 years. The pavilions and terraces were left unrepaired; military occupation has hastened their decay. The opera stage, once vibrant, and the flower beds now only hint at the past splendor of the garden. With the family having relocated to Shanghai long ago, visitors cannot help but feel the melancholy and abandonment, much like the willow trees that have outlived their time."

In 1944, the garden was sold to Yin Rugeng (1883–1947), a notorious collaborator with the Japanese invaders. This marked the end of an era, as the garden, which had been the ancestral home of the Ho Family for 61 years, ceased to be a place of their glory and memories.

归隐寄啸为主题营造了北部大花园的山水空间。在此基础上，他还在各个院落和建筑之间设计和建造了复道回廊，并在东南角新筑了家祠，从而构成寄啸山庄的整体面貌。园林建成以后，因其巧妙的构思、精美的建筑、别致的山水空间，被时人赞誉为“咸同后扬州城内第一名园”。

何维键在何园居住了18个春秋，至光绪二十七年(1901年)，何氏举家南迁上海，投资大清银行、汉冶萍煤铁厂矿公司等实业，何园则交由管家看管。光绪三十四年（1908年），何维键辞世，享年73岁。1935年10月，为节省生活开支，长子何声灏一家回到何园居住。然而，1937年，抗日战争全面爆发后，何声灏全家祖孙三代20余人再次离开何园，辗转返回上海。这一时期，著名园林专家童寯先生曾到何园考察，后在其著作《江南园林志》中写道：“何园为扬州私园之最大而仍存者。住宅在东部，中部正厅，院落重重。西有洋式房屋三排，其后即园林部分，称寄啸山庄。园为盐官何芷舠所建，垂五十年，亭台失修，益以驻军，荒圮日盛。山池之外，为戏台花圃，徒见昔日梨园药栏之盛，主人久已移居沪滨，游者不禁生柳老堂空之感。”1944年，何氏后代将寄啸山庄出售给汉奸殷汝耕，这座承载了何家整整61年荣耀与记忆的园林，至此画下了作为何家住宅的句点。

## Research and Education Base

The three decades following the change of ownership of the garden coincided precisely with the founding of the People's Republic of China and the critical period of national economic recovery. During this period, the garden underwent a remarkable transformation—from a private estate to serving successively as a military facility, research, and educational base, making significant contributions to both regional development and national reconstruction.

Following the victory in the Chinese People's War of Resistance Against Japanese Aggression in 1945, the former Nationalist Government requisitioned the Ho Family Garden as enemy-collaborator property. Over the next four years, the estate served as the temporary campus for Zhutong High School when it was relocated to Yangzhou. After 1949, the compound was successively utilized by military and government entities, including the Northern Jiangsu Military District Command.

In 1959, the landscape section north of the Whistling Scholar's Mountain Retreat was transferred to the Yangzhou Municipal Landscape Administration and was opened to the public following a restoration. Three years later, Chen Congzhou conducted extensive fieldwork on the gardens and historic architecture in Yangzhou. During his research, he discovered the

# 科教驻地

何园易主后的三十余年，恰恰是中华人民共和国成立和国民经济恢复的重要历史时期，何园从一座私家花园迅速转变为军事、科研和教育的场所，为国家和地方的发展与建设作出了重要贡献。1945 年抗日战争胜利以后，国民党政府将寄啸山庄作为敌伪财产接收，之后的 4 年间，此处借给迁入扬州的祝同中学上课。中华人民共和国成立后，何园先后为苏北军区等单位所驻用。1959 年，扬州市人民政府将寄啸山庄北部园林部分移交给市园林部门，经过维修后对外开放。三年后，陈从周在扬州开展园林与古建筑调查期间，发现了片石山房假山遗存，他从堆叠手法的精妙、形制的古朴着手，再证以史料，推断片石山房假山为石涛堆叠，并呼吁对这一“石涛叠山的人间孤本”开展研究和保护。1963 年，何园后花园及祠堂以外的部分交给国防部第七研究所驻用，后又相继为六机部七院第五研究所、六机部七院第十研究所驻用。1969 年，扬州市革命委员会曾将园林部分交由扬州无线电厂使用，忙碌的生产活动不免“误伤”园内的山水花木。1972 年，六机部七院第二十三研究所搬迁至此，将何园的住宅部分和片石山房部分作为办公场地，直到 20 世纪 80 年代迁出。

historic remains of the Small-Rock Mountain Retreat and verified that the rockery was an artifact created by the grandmaster Shi Tao himself. His conclusions were based on a meticulous review of the sophisticated stonework techniques, archaic configuration, and relevant historical documents. Chen then advocated for further research and conservation measures for this cultural relic, describing it as "the only surviving example of rockery formation by Shi Tao."

In 1963, the rear garden and the area outside the ancestral hall were designated for use by the 7th Academy of the Ministry of National Defense. The site later hosted Research Institutes 5 and 10 from the 7th Academy of the Sixth Ministry of Machine-Building. By 1969, the garden section was transferred to the Yangzhou Radio Factory by the Yangzhou Revolutionary Committee. Unfortunately, the factory's bustling industrial operations inadvertently caused damage to the landscape features, including the rockeries and historic plantings. In 1972, Research Institute 23 moved to the site and used both the residential quarters and the Small-Rock Mountain Retreat as office spaces until the 1980s.

片石山房内院入口
Entrance to the Small-Rock Mountain Retreat

# Cultural Heritage

In 1979, the northern section of the Ho Family Garden, previously occupied by the Yangzhou Radio Factory, was returned to the municipal landscape department. Shortly thereafter, the garden opened to the public following extensive restoration of its rockeries, meandering pools, towers, pavilions, belvederes, and plantings. In 1985, the residential quarters, along with the Small-Rock Mountain Retreat, which had been occupied by Research Institute 23, were transferred to the Yangzhou Municipal Landscape Administration. On January 13, 1988, the Whistling Scholar's Mountain Retreat was designated as one of the third batch of the Major Historical and Cultural Sites Protected at the National Key Level, cementing its status as a treasure of national architectural heritage. In April 1989, the expert Wu Zhaozhao was tasked with overseeing the restoration of the rockery at Small-Rock Mountain Retreat, aiming to restore the original layout designed by Shi Tao, with a particular emphasis on the connection of its western peak and eastern fragments.

To the southeast, a series of lower caves and crevices contrasted sharply with the towering peaks. A gazebo was built on the southern bank of the pond, extending over the water and facing the rock formations on the opposite side. To the east, serpentine walkways led to the Nanmu Hall, while a hip-and-gable-roofed land boat (*xieshan fangwu*) was positioned

# 文化遗产

扬州无线电厂于1979年搬出，何园北部复归园林部门管理，逐步修复假山、曲池、厅堂楼阁，补植花木，并重新向游客敞开了大门。到了1985年，何园的住宅部分及片石山房由六机部七院第二十三研究所移交给扬州园林局。1988年1月13日，寄啸山庄（何园）被国务院公布为第三批全国重点文物保护单位，其国家级文化遗产地位正式确立。次年4月，园林部门组织修复片石山房，委托园林专家吴肇钊进行设计，将残存的西侧主峰和东侧片段按石涛布局章法重新连接修补，在东南侧堆叠水岫洞壑，以山洞之幽深衬托石峰之峻峭，水池南岸建造水榭横跨水面，与假山隔水相望。东侧建曲廊与楠木厅相接，楠木厅西山墙建造歇山舫屋。同时，将池水引入水榭围栏，并设置琴台、棋墩、书房、花窗等元素，进一步丰富了景观。

四十余年来，扬州市园林管理部门始终致力于何园的保护与修复。2002年，玉绣楼北楼及何家史料馆率先对外开放，2003年，东二楼和东三楼完成修复，玉绣楼南楼随之对外开放；2004年，骑马楼得以恢复，使全园复道回廊全线贯通，较好地再现了晚清扬州私家园林的风貌；2005年，东大门南墙壁上的“晚清第一园”石刻翻建完工，这一题字由古建专家罗哲文先生所书，为何园再添文化韵味；随着何家祠堂于2007年收回整修、回归公众视野，何园的主要景观至此基本复原，逐渐呈现出今日所见的面貌。

against its western wall. The water of the pond flowed beneath the balustrades of the gazebo, and features such as a *qin* table, a chess table, a study studio, and latticed windows were added, enhancing the garden's visual and cultural appeal.

For over forty years, the Yangzhou Municipal Landscape Administration has dedicated itself to the preservation and restoration of the Ho Family Garden. In 2002, the restoration project achieved its first major milestone with the public opening of the northern section of the Yuxiu Building and the Ho Family Historical Exhibition Hall. The following year, in 2003, restoration efforts continued with the completion of Building 2 and 3 in the East Wing, along with the unveiling of the southern part of Yuxiu Building. A significant moment occurred in 2004 with the reconstruction of the Horseback Riding Building, which allowed for the full interconnection of the multi-level walkways. In 2005, the stone inscription on the southern wall at the east gate was finalized. This inscription, written by Luo Zhewen (1924–2012), a grandmaster of classical Chinese architecture, reads "the Best Garden of the Late Qing Dynasty." Finally, in 2007, the restoration of the Ho family ancestral hall was completed, marking the conclusion of major landscape restoration in the garden and making it accessible to the public.

何园修缮工程记录，*2016* 年
Record of the 2016 restoration of Ho Family Garden

# 游园

ROUTES

## 楼阁玲珑合四环

Elegant Buildings and Exquisite Belvederes in Fourfold Harmony

The Ho Family Garden exemplifies the Jiangnan tradition of combining living spaces with gardens—an ideal balance of residence and leisure. Its layout is defined by a sequence of interlinked courtyards, seamlessly integrated with gardens of varying scales and styles, forming a poetic fusion of architecture and nature.

Upon entering, visitors are greeted by the grandeur of a secluded mansion: two-story lodgings, graceful pavilions, and a fourfold courtyard layout that conveys elegance and order. As one ventures deeper, the garden reveals a different charm—artfully arranged rockeries, ponds, flowers, and trees create layered, ever-changing scenery. Buildings and landscapes are interwoven to offer a sense of refinement without overcrowding, and intimacy without constraint.

Strolling through the Ho Family Garden is like reading a well-structured essay, with transitions that unfold new perspectives at every turn. Each courtyard bears its own unique composition and historical significance, and together they contribute to an immersive and resonant spatial experience.

Today, the garden comprises four main areas: the Little Garden, the Ho Family Ancestral Hall, the residential courtyards, and the Grand Garden. With dozens of scenic sites and more than a hundred unique features, every visit yields a different impression, shaped by season, weather, and mood. Based on historical context and the present layout, we propose a route from the South Gate on Huayuan Lane to the rockeries at the West Garden, highlighting ten key scenic sites that best reflect the garden's aesthetic beauty and cultural significance.

江南宅第园林多是院落和花园的组合，院为居，园为游，古人在居与游的交错和平衡中，营造出理想的生活场所。何园以建筑院落组合构成整体空间的骨架，在其中回环穿插不同规模、不同形式的花园，构成人工与自然交错的园居景观。初入何园，第一印象是楼阁玲珑、四环合围的深宅大院，颇具富丽堂皇之感。但走进院内细观，分布其间的山池花木构思精巧、清新雅致，与亭台楼阁相互掩映，景物虽密但不觉繁复，院落虽多但不显拥塞，给人以步移景异、层层递进的感受。

游览园林如同阅读文章，在特定线路中追求起承转合的变化。何园由于其丰富的历史层积和多样的空间内涵，不仅每个院落单元都有其独立的景观结构，相互组合之后更是产生了无穷的意蕴。今日的何园，按功能可以分为小花园、何家祠堂、园居院落、大花园四个部分，其中包含景点几十处、景物百余个。因此，可游的线路不计其数，景观感受更是因时、因人而异。结合历史视角和园林现状，我们设定了一条自花园巷南门开始，至西园假山结束的游线，并甄选出最具代表性的十景，一方面解读景观的特征，另一方面阐释其蕴含的历史典故和文化，期待能较为全面地呈现何园的景象与境界。

| | | |
|---|---|---|
| 01 | 片石山房 | Small-Rock Mountain Retreat |
| 02 | 何家祠堂 | Ho Family Ancestral Hall |
| 03 | 玉绣楼 | Yuxiu Building |
| 04 | 赏月楼 | Moon-Viewing Building |
| 05 | 贴壁假山 | Wall-Backed Rockery |
| 06 | 船厅 | Land Boat Hall |
| 07 | 水心亭 | Water-Heart Pavilion |
| 08 | 蝴蝶厅 | Butterfly Hall |
| 09 | 复道回廊 | Multi-Level Walkway |
| 10 | 西园假山 | Rockeries at the West Garden |

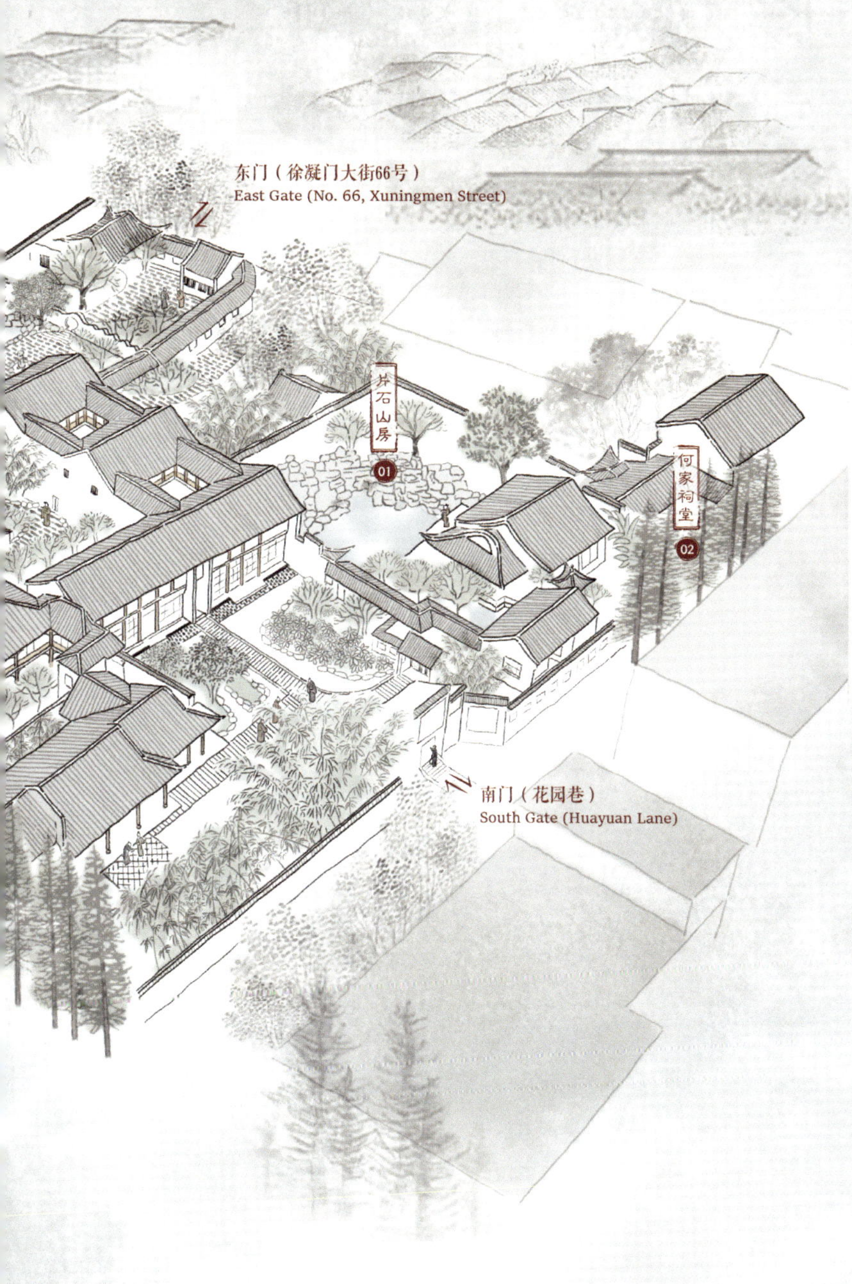
东门（徐凝门大街66号）
East Gate (No. 66, Xuningmen Street)
片石山房
01
何家祠堂
02
南门（花园巷）
South Gate (Huayuan Lane)

Entering the garden through the South Gate, one is greeted by the Small-Rock Mountain Retreat, a charming retreat once known as the "Little Garden." This area carries forward the legacy of Wu Jialong's Twin Sophora Garden, upon which it was built. Over the years, it has become the most historically significant site in the garden. In the entrance atrium stands "One-Drop Spring" (*Yidiquan*), a miniature water feature inspired by the concept of "raindrops forming a waterfall." The whitewashed wall bears a stone inscription titled "Record of the Restoration of the Small-Rock Mountain Retreat," written by Chen Congzhou. This inscription describes the discovery and restoration of this historic relic.

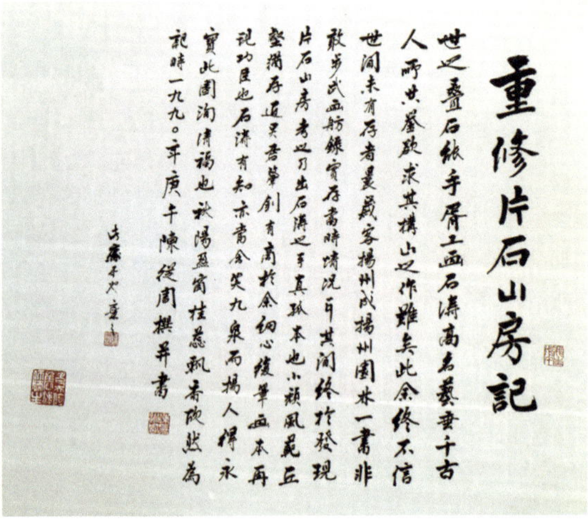


《重修片石山房记》原作
Original calligraphy of "Record of the Restoration of the Small-Rock Mountain Retreat"

从南门步入何园，着目即是当年被何家人称为“小花园”的片石山房，这座精美的花园承袭了吴家龙双槐园的遗迹，并在岁月流转中逐步演变，成为了何园最具历史价值的部分。进入片石山房，门厅天井有“一滴泉”——以“注雨观瀑”为主题的小水景，粉墙上嵌陈从周先生所撰《重修片石山房记》石刻碑文，记录了当年发现和修复片石山房的过程。

《重修片石山房记》

世之叠石能手胥工画，石涛高名，艺垂千秋，人所共鉴。欲求其构山之作，难矣。然余不信世间未有存者。曩岁客扬州，成《扬州园林》一书，非敢步武《画舫录》，留真况耳。其时终于发现片石山房，考之，乃出石涛之手，孤本也。小颓风范，丘壑犹存。近吴君肇钊就商于余，细心复笔，画本再全，功臣也。石涛有知，亦当含笑九泉，而扬人得永宝此园，洵清福，无量矣。

——一九九零年庚午，陈从周撰并书

《重修片石山房记》
"Record of the Restoration of the Small-Rock Mountain Retreat"

片石山房鸟瞰
Aerial view of the Small-Rock Mountain Retreat

# Scene 1: Rockeries at the Small-Rock Mountain Retreat

As one enters the entrance hall, the Small-Rock Mountain Retreat gradually reveals itself. At its center is an excavated pond, against the northern wall of which stretches a horizontally arranged Lake Tai limestone rockery—a crowning treasure of Ho Family Garden. The term "Small-Rock" refers to the collection of small Lake Tai rocks that comprise this rockery, while "Mountain Retreat" suggests that it mimics the appearance of multiple chamber-like caves within. The westernmost section of the rockery features a prominent peak, distinguished by its elegant, jagged rock formations. To the east of this main peak, the rockery includes lower caves and crevices that create shadowed recesses, enhancing the dramatic effect of the steep cliffs. This design creates a harmonious interplay of depth and height.

At the slightly lower central area of the rockery, a winding path leads southeast, lending a sense of timeless solidity. Though modest in scale, the rockery is artfully designed, balancing density and openness. Its composition follows the principle of "one peak rising abruptly, followed by ridges and ravines," echoing a painting philosophy where "peaks and texture strokes unite."

This rockery is noted for three remarkable features.

# 第1景 片石山房假山

穿过门厅，片石山房花园的主要院落徐徐展开，院内中部凿池，池北依墙堆叠一座横长形的太湖石假山，这便是何园的镇园之宝——片石山房假山。所谓“片石”，是指这座假山皆以小块太湖石拼镶而成；所谓“山房”，则是指假山内部别有洞天。假山贴北墙而筑，西首为主峰，奇峭秀逸，片石峥嵘。主峰之东，山石层叠，构成低伏幽深的水岫洞壑，以曲折隐宕衬托石峰的峻拔峭丽。中部山体略低，但古朴浑厚，沿着山道逶迤向东南行，一直延伸至园林东墙。此山占地面积不大，但布置自然、疏密适当，具有“一峰突起，连冈断堑，变幻顷刻，似续不续”的章法，形态符合“峰与皴合，皴自峰生”的画理。

片石山房假山有三绝可观：其一，主峰特立耸秀，奇峭夭矫，是我国江南园林现存太湖石假山中最具风姿的片段之一，峰前有高梧映带，山中有磴道可攀，曲干虬枝更加强化了峰峦的俊逸多姿；其二，在主峰腹中以青砖砌筑两间小屋，这是“山房”之名的奥秘所在，夏日入内，暑热尽消，可尽享山居悠然、隐逸宁静之乐；其三，山间种有一株小叶罗汉松，树龄已有340余年，据推测乃是吴家龙双槐园时期所植，极为珍贵。

First, it boasts a soaring main peak, showcasing an elegantly rugged form. With its dynamic curves and striking presence, it is one of the most distinguished examples of Lake Tai rockeries in Jiangnan gardens. Second, cleverly hidden within the belly of the main peak lie two small chambers built of traditional blue-gray bricks—this architectural secret reveals the origin of the name "Mountain Retreat." In the summer, these secluded spaces remain refreshingly cool, providing a tranquil

据考证，这座假山出自我国明末清初的画坛巨匠石涛之手，他原名朱若极，是扬州画派的开创者，也是扬州八怪的先驱。石涛遍访名山大川，正所谓“搜尽奇峰打草稿”，晚年侨居扬州，参与当地的文化艺术和园林创作活动。石涛曾写过一首诗：“白云迷古洞，流水心澹然。半壁好书屋，知是隐真仙。”生动反映了片石山房假山的意境追求。

片石山房假山

Rockeries at the Small-Rock Mountain Retreat

retreat where visitors can enjoy the serenity of mountain living and the charm of the retreat. Lastly, nestled within the rockery is a small-leaved Buddhist pine (*Podocarpus brevifolius*) over 340 years old. It is believed to have been planted during the time of Wu Jialong's Twin Sophora Garden, making it a rare and precious botanical treasure.

Art historical evidence attributes this rockery to Shi Tao, originally named Zhu Ruoji. He was a foundational figure in the Yangzhou School of painting and a precursor to the "Eight Eccentrics of Yangzhou." Throughout his life, Shi Tao traveled extensively across famous mountains and rivers, embodying his artistic creed: "To seek out all wondrous peaks to draft your sketches." In his later years, he settled in Yangzhou, where he actively participated in the cultural, artistic, and garden design circles. He once penned a poem that reflects his philosophy:

*White clouds veil ancient caves,*
*While flowing water empties all worldly cares.*
*A study carved halfway up the cliff—*
*Here dwells the truth-seeker,*
*beyond mortal affairs...*

This verse vividly captures the spiritual pursuit represented by the Small-Rock Mountain Retreat.

Across the pond stands Nanmu Hall, likely from the mid-to-late Ming dynasty and the oldest surviving timber-frame building in Yangzhou's gardens. Facing

水池以南为楠木厅，与片石山房假山隔水相望，应是明代中后期遗构，也是扬州园林中现存最古老的木构建筑。楠木厅坐北朝南，面阔三间，硬山屋顶配小瓦屋面，出檐深远。木构为抬梁式架构，肥梁硕柱，梁架不用漆面，纯以材质本色取胜，整体风格简洁古朴，浑厚端庄。楠木厅东为水榭，三楹面北，中置泉井，西室建有半壁书屋，中室涌泉伴有琴桌，东室有棋台一块，造型古拙，乃是1979年双槐园两棵老槐树因病被砍去之后利用树桩制作而成，是重要的历史见证。东室抬头可见竹石图，颇有琴、棋、书、画兼备的雅趣。楠木厅西侧有一“不系舟”舫屋，坐在舟上可俯视池鱼，仿佛船行水中。园内西廊壁上置有碑刻，选用石涛诗文等九篇，壁上还嵌置一块硕大镜面，使得整个园景可通过不同角度映照其中。陈从周曾在《园韵》中著文，描述片石山房部分“逶迤衡直，闿爽深密，曲具中国园林的特征，为这时期的代表作品”。

south, it features a three-bay façade, a hard-hill roof with clay tiles, and broad eaves. Its unvarnished nanmu beams and columns, highlighting natural wood grain, reflect the solemn, unadorned grandeur typical of Ming-era architecture.

To the east of Nanmu Hall, there is a waterside gazebo featuring three north-facing bays that center around a stone-rimmed well. The western chamber is home to the "Half-Wall Study Studio," while the central area showcases a spring-fed pool, which is flanked by a *qin* table. In the eastern chamber, there is a Go (*weiqi*) board crafted in 1979 from the diseased twin sophora trees, serving as a poignant reminder of the garden's history. The ceiling is adorned with bamboo-and-rock paintings, completing the quartet of scholarly arts: *qin*, Go, literature, and painting. To the west, a Free-Moored (*buxizhou*) land boat extends over the pond, allowing for views of koi swimming beneath its cantilevered floor. The wall of the western gallery displays nine engraved poems by Shi Tao, while a strategically placed mirror enhances the spatial illusions of the garden.

As noted by Chen Congzhou in *The Poetics of Chinese Gardens*, "Here in the Small-Rock Mountain Retreat, the dance between serpentine trails and geometric precision, between airy clarity and verdant mystery, manifests the very soul of classical Chinese garden design—a definitive masterpiece of its epoch."

片石山房楠木厅（上）与半亭（下）
Nanmu Hall (top) and Half-Pavilion (bottom) at the Small-Rock Mountain Retreat

As you walk east along the southern wall of the Small-Rock Mountain Retreat, you will arrive at the Ho Family Ancestral Hall. This ancestral hall spans over 600 square meters and comprises several single-story and brick-timber structures, and one two-story main structure. The great hall, known as the Hall of Illuminating Virtue (*Guangde Tang*), is named after an admonition from He Jun (1797–1858), the father of garden owner He Weijian. He Jun taught that "when entering the ancestral hall, one may commune with departed spirits; by studying family records, one can understand and illuminate the virtuous legacy of our forebears."

This complex consists of two ceremonial halls arranged in an east-west orientation, each with its own courtyard, creating a particularly elegant design for such a structure. The western hall, known as the Ancestral Worship Hall (*Xiang Tang*), features an adjoining side chamber that once functioned as the family accounting office for managing financial records. Its walls are decorated with the Ho Family Precepts, which are a collection of moral teachings. The eastern hall, known as the Hall of Repose (*Qin Tang*), functions as the principal ceremonial space, housing remarkably preserved portraits of five generations of Ho ancestors. At its center lies an ancient well, pragmatically providing water for rituals while symbolically reminding descendants to "remember the source while drinking"—a perpetual exhortation to honor their lineage's legacy.

沿片石山房院落南墙向东行，便到了何家祠堂，这里原为何家供奉祖先牌位的地方。现存的祠堂占地面积为600多平方米，包含多间砖木结构的平房和一座两层主楼，正厅名光德堂，“光德”取自园主何维键之父何俊“登祖宗之堂可对先灵读传记之文，可光旧德我”的训喻。祠堂院内有东西一字排开的两座厅堂，各有院落，十分讲究，是不多见的祠堂范例。西为享堂，堂西带有一厢房，是签押房，当年何家用作管理各种往来账目之处。享堂四面墙上悬挂《何氏家训》。东为寝堂，是祠堂的正厅，供奉有保存完好的何氏五代祖宗容像。厅内有一口古井，既方便了祭祀用水，又能在家祭时提醒子孙饮水思源，不忘祖宗恩情。

何家祠堂入口
Entrance of the Ho Family Ancestral Hall

## Scene 2: Ho Family Precepts

The ancestors of the Ho clan once established sixteen lineage regulations and eleven family rules to guide their descendants. These guidelines emphasized filial piety towards parents, respect for elders and superiors, the friendly relationship with neighbors, proper discipline and education of children, the pursuit of honorable livelihoods, and the avoidance of wrongdoing. These principles were intended to uphold moral conduct and ensure the family's enduring virtue. One important instruction stated that "diligence refines one's profession, while idleness results in its decline." This applies to all occupations—be it scholar, farmer, artisan, or merchant—each must be pursued with earnest dedication. Additionally, the precepts advocate for frugality, emphasizing that life's blessings are found in moderation. One can maintain enduring prosperity by practicing thrift in food, clothing, expenses, and daily living. Education was highly valued in the Ho family. One key rule indicated that "to determine whether a child will succeed in learning, one need not examine their temperament or intelligence, but rather observe their sense of respect—this alone reveals the breadth of their future vision." The elders instructed that learning should not be pursued for fame or fortune; instead, individuals must intensively study according to their aptitude and capabilities. Only in this way

# 第 2 景 何氏家训

何氏先人曾经撰写“宗规”十六条以及“家规”十一则，教育子孙后代须孝顺父母、尊敬长上、和睦邻里、教训子孙、各安生理、毋作非为。家训中提到，“勤则职业修，惰则职业惰”，士农工商所从事的职业虽然不同，但都是本职工作，都需要认真对待。家训还教育后人要崇尚节俭，人生的福分在于有所限制，如果能够在吃穿、用度、起居上都做到节俭，则可以保持富余，并享之不尽。何氏家族极为看重子女教育，家训中撰文：“欲知子弟读书之成否，不必观其气质，亦不必观其才华，先要观其敬与不敬，则一生之视野概可见矣。”先辈教导子女读书者不应以富贵功名为目标，而应根据自己的能力，专心向学，方能成为有用之才。

这些重要的人生信条，为何氏家族自元代以来长时期的兴旺与成就提供了精神上的有力支持。园主何维键的父亲何俊（1797—1858），道光己丑年（1829 年）以三甲第一名进士被选为翰林院庶吉士，清代道光至咸丰年间，历任南河海阜海防同知，广西桂林府知府、直隶大顺广道道台、江苏布政使，成为治理一方的贤才；何维键（1835—1908）在湖北汉黄德道兼江汉关监督任上，

can they truly become valuable talents and men of virtue.

These core values of the Ho Family have provided enduring spiritual support for their prosperity and achievements over the centuries since the Yuan dynasty.

He Jun, father of estate owner He Weijian, distinguished himself by ranking first in the third tier of the 1829 Imperial Examinations. He was appointed as a Probationary Compiler at the Hanlin Academy and later served as the subprefect of Nanhe, Haifu, and Haifang, as well as the prefect of Guilin in Guangxi and circuit intendant of Da-Shun-Guang area in Zhili Province. His governance earned him a reputation as a capable and virtuous official.

His son, He Weijian (1835–1908), further upheld the family's legacy. While serving as the circuit intendant of Han-Huang-De area in Hubei Province and the superintendent of Hankou Customs, he defended national sovereignty, resisted foreign pressure, and contributed to economic and social development as well as public welfare, demonstrating exceptional administrative talent.

He Shenghao (1864–1941), the eldest son of He Weijian, achieved the degree of Jinshi in 1890, ranking the sixth in the second class of the Imperial Examination and subsequently earned first-class honors in the Court Examination. Appointed as a probationer at Hanlin Academy by Emporor Guangxu, he later became the director of the Hanyeping Coal and Iron Company, which

不畏强权，不谀外、不媚上，维护国家主权，并且在民生事务和经济发展方面展示了出众的才能；何维键长子何声灏（1864—1941），于清光绪十六年殿试二甲第六名进士，朝考一等，钦点翰林院庶吉士，并曾作为张之洞创办的汉冶萍煤铁厂矿公司董事进行实业救国的探索，还曾出资支持孙中山先生的革命事业；何维键孙子何世桢、何世枚二人均毕业于东吴大学，并获得美国密歇根大学法学博士学位，被称作“兄弟博士”，后回国开设法律事务所，被誉为中国海洋法系的奠基人，二人于1924年创办的持志大学，是中国高等教育史上最早的私立大学之一；何声灏次子何适斋（1893—1985）为书画名家，被国画大师黄宾虹赞誉为“沉浸六朝，清劲拔俗”，他与黄宾虹等人共同发起和组建了中国金石书画艺观学会。从何家走出的杰出人物还有中国最早进入联合国工作的人员之一何祚藩（1913—2013），享受国务院特殊贡献津贴的资深翻译家何祚康（1928—2015），中国科学院院士、为我国第一颗原子弹研制作出突出贡献的王承书（1912—1994，何声灏外孙女），中国科学院院士、理论物理学家何祚庥（1927—）等。

was founded by Zhang Zhidong. This role placed him at the forefront of the late-Qing "industrial salvation" movement. Notably, he also provided financial support to Dr. Sun Yat-sen's revolutionary cause.

He Shizhen and He Shimei, grandson of He Weijian, both earned law degrees from Soochow University and the University of Michigan. Known as the "Doctor Brothers," they played a significant role in establishing the Common-Law school within China's legal community and founded Chizhi University in 1924, one of China's first private universities.

He Shizhai (1893–1985), the second son of He Shenghao, was a renowned painter and calligrapher. He was highly praised by the master Huang Binhong, who described him as someone who "immersed himself in the Six Dynasties' tradition, achieving a style of pristine vigor and transcendent elegance." Alongside Huang Binhong and other prominent artists, He Shizhai co-founded the Chinese Society for the Study of Stone Inscriptions, Calligraphy, and Painting.

Other notable family members include: He Zuofan (1913–2013), one of the first Chinese professionals to work at the United Nations, He Zuokang (1928–2015), a senior translator who received special government allowances from the State Council, Wang Chengshu (1912–1994), the granddaughter of He Shenghao, a scientist who contributed to China's first atomic bomb, He Zuoxiu (1927– ), a theoretical physicist and academician of the Chinese Academy of Sciences.

何祚宏二弟兄（上），何祚惠与王承礼（下）在何园，1935 年摄
Three He brothers (top) and He Zuohui with Wang Chengli (bottom) at Ho Family Garden, 1935.

The residence quarter in the Ho Family Garden, which extends westward from the ancestral hall, serves as the primary living and reception area. The architectural complex features a blend of Chinese and Western styles and comprises two parallel axes running east and west. Along the main western axis, from south to north, you will find the Warm Spring Hall, the Yuxiu Building, and the Moon-Viewing Building. The secondary eastern axis includes the Horseback Riding Building, as well as Building 2 and 3 in the East Wing. This layout not only highlights the grandeur of the Ho Family's expansive courtyard residence but also showcases the diversity of its architectural styles.

The Warm Spring Hall, situated at the southernmost part of the garden, is also known as the Hall of Communal Retreat (*Yugui Tang*). Constructed from precious golden nanmu wood, it is often referred to as the "Nanmu Hall." Among the classical gardens in Yangzhou, it is the youngest but the best-preserved and largest surviving nanmu hall. It serves as the main reception room for the Ho family, where the host would formally entertain guests. It is believed that during the residence of the Ho family, the primary entrance to the garden was situated south of the hall.

Its great hall is architecturally impressive, featuring seven bays in width and a single-eave hip-and-gable roof, with a raised central section over the three main bays. It has a spacious footprint of over 100 square meters, exuding an atmosphere of elegant grandeur and openness. On the front roofed walkway, there are columns that display a couplet written in cursive script

自祠堂西行，铺陈开的是何园的住宅区域，这里也是当年园主人起居和待客的主要场所。这座中西合璧的院落，以东西两条轴线串联，西侧主轴线上自南向北有煦春堂、玉绣楼、赏月楼，东侧次轴线上有骑马楼、东二楼、东三楼等，突出展示了何园深宅大院的气势及多样化的建筑风格。

煦春堂坐落在全园最南面，又名与归堂，因堂内木构件都采用金丝楠木，故又称“楠木厅”，属扬州园林中最年轻，也属现今保存最完好、面积最大的楠木厅。这里是何家宅园的正门客厅，即园主接待宾客的正式场所。据推测，何氏家族居住时期，何园的主入口位于煦春堂南侧。煦春堂面阔七间，单檐歇山顶，中间三间正厅屋顶抬高，占地面积达一百多平方米，十分气派、舒朗。前廊柱上悬挂何绍基行书对联“退士一生藜苋食，散人万里江湖天”，堂内金柱上配联“莫放春秋佳日过，最难风雨故人来”，点明该建筑的主旨。值得一提的是，煦春堂大厅的门窗全部采用进口玻璃装饰，使得室内更加敞亮，彰显了园主人当时雄厚的财力。

何园深宅大院的格局与多样建筑风格
The grand layout and diverse architectural styles of the Ho Family Garden

by the renowned calligrapher He Shaoji (1799–1873), which reads that "A recluse sustains on wild herbs; / A wanderer roams rivers and lakes under vast skies." Inside the hall, the central pillars bear another couplet that reads: "Let not the fair days of spring and autumn slip by; / How blessed, through rain and wind, old friends are nigh." These inscriptions poetically capture the essence of the great hall, celebrating reclusive elegance and enduring friendship. Notably, its windows and doors are adorned with imported glass. This luxurious feature floods the interior with light and serves as a striking testament to the owner's wealth and modern tastes at that time.

玉绣楼内院
Inner courtyard of Yuxiu Building

## Scene 3: Yuxiu Building

To the north of The Warm Spring Hall is a traditional quadrangle courtyard compound known as the Yuxiu Building, which was named after two plants growing in the courtyard, Magnolia grandiflora (*yulan*) and Hydrangea macrophylla (*xiuqiuhua*) growing in the courtyard. The name, Yuxiu, was given by Li Shenghe (1908–2001), a renowned painter and calligrapher, in the 1990s. This two-story structure served as the primary residential quarters for the Ho family and was nicknamed the "Western-Style Building" (*Xiao Yanglou*) due to its unique architectural blend of Chinese and European elements. This compound is an excellent example of large-scale residential architecture from the Qing era, featuring a thoughtfully designed rectangular layout that showcases superior material selection, form, and craftsmanship. It consists of two two-story brick-and-timber wings located north and south, which are connected by roofed walkways in the center. This design merges traditional Chinese *chuanlou* (linking gallery) concepts with Western architectural styles. Each wing contains six rooms on each floor, totaling twenty-four rooms overall. The layout is highly functional and well-organized, with every three-room unit designed around a central doorway that leads to a staircase flanked by two side rooms.

# 第 3 景 玉绣楼

煦春堂以北是一处四合院落，20 世纪 90 年代，书画家李圣和先生将此楼命名为玉绣楼，取楼院中栽植的广玉兰和绣球之名。这里是当年园主人主要的居住空间，在何氏居住时期被唤作“小洋楼”。该建筑呈长方形环合布局，在用材、造型及营造技艺等方面体现出高度的实用性与艺术性，堪称清代大型民居建筑的代表性作品。玉绣楼包含南北两座砖木结构的二层楼，中部由楼廊连接。玉绣楼的建筑设计既采用中国传统式的串楼理念，又融入西方建筑手法。南北楼上下两层各 6 间，共 24 间，每 3 间一套，每套的中门都是楼梯间，左右两边为房间，布局十分合理。

玉绣楼南楼
South wing of Yuxiu Building

The building features a ventilated substructure with air vents integrated into the flooring, demonstrating a thoughtful approach to both scientific and comfortable residential design. It is spacious and elegantly designed, utilizing high-quality stone materials and terrazzo bricks. In the first exhibition hall of the north wing's archive room, a calligraphy couplet is displayed: "One curtain of wind and moon, like a painting by Wang Wei. / Four walls of clouds and mountains, reminiscent of the poems by Du Fu." This couplet is believed to have been composed by the Qing scholar Sun Xingyan (1753–1818), and the calligraphy itself is the work of He Shizhai, the grandson of He Weijian. The structure incorporates a variety of architectural decorative elements from different cultures, featuring European-style railings, arched carved brackets, Japanese sliding doors, adjustable louvered shutters, Western chandeliers and fireplaces, wrought-iron beds, and various other furnishings. These innovative design choices were considered highly modern for their time, reflecting the garden owner's extensive overseas experiences and his open-mindedness for the alien.

建筑地下有透气层，地面设通风孔，充分考虑了住宅建筑的科学性和舒适性。玉绣楼整体华丽轩敞，用石精细，砖皆水磨。北楼资料陈列室第一展厅悬挂行书对联：“一帘风月王维画，四壁云山杜甫诗”，或为清代学者孙星衍所写，书法为园主何维键之孙何适斋笔。玉绣楼杂糅了国外的建筑装饰风格，运用了洋式栏杆、拱形挂落、日式拉门、活动遮阳百叶窗、西式吊灯和壁炉、铁艺床等，独具匠心，在当时是非常新潮的家居设施，反映了园主出洋积累的丰富阅历和万物为我所用的开放思想。

玉绣楼院子中的木绣球
Chinese snowball viburnum in the courtyard of the Yuxiu Building

## Scene 4: Moon-Viewing Building

To the west of the Yuxiu Building is the Moon-Viewing Building (*Shangyue Lou*), also known as the Serenity Building, which He Weijian built for his mother, Madam Ren. This structure connects to the rear garden to the north and the Yuxiu Building to the east. It faces south and consists of three bays across two stories, featuring a hip-and-gable roof on the west side. Small doors on both floors connect the building to other residential buildings, and a multi-level walkway provides convenient access between different sections. Its design emphasizes themes of filial piety and good fortune, adorned with iron railings inscribed with the phrases "Longevity and Good Health." Currently, it serves as an exhibition hall displaying the Ho family's genealogical records and ancestral portraits. A rockery with a winding stone path in the front represents "Longevity as Eternal as the Southern Mountains." The space features various plants, including pines symbolizing longevity, privet trees representing integrity, crape myrtles for harmony, pomegranate trees signifying prosperity and fertility. The courtyard is adorned with cobblestone patterns that embody fortune, prosperity, longevity, and happiness, reflecting He Weijian's dedication to filial piety. During

# 第4景 赏月楼

玉绣楼隔墙之西为赏月楼，这里是当年何维键为生母任太夫人所建的居所，北与后花园相连接，东与玉绣楼住宅相连。主楼赏月楼又名“怡萱楼”，坐北朝南，三楹两层，西部为歇山顶。赏月楼上下，各有小门与住宅楼相通，且与复道回廊相接，交通十分便捷。赏月楼以敬孝、吉祥为主题，外廊上铁栏花案有“延年益寿”字样，清静幽雅，现已辟为陈列室，陈列着何氏家族世系谱牒及近代有功名祖先礼服容像；楼前小院中叠湖石山，有“寿比南山”的寓意，山有腹洞，有磴道盘折达于楼上西廊；山中栽植有松柏、女贞、桂树、紫薇、石榴等花木，柏象征长寿，女贞象征节操，紫薇象征和睦，石榴象征多子；楼前地面也用鹅卵石铺成福、禄、寿、喜图案，反映了园主对生母的孝心，每逢中秋金桂飘香之时，这里就变成登楼赏月的佳处。

玉绣楼隔墙之东为骑马楼，是当年宅园的客房，两者之间有火巷相隔。“骑马”之名寓异乡与征途，别有一番深意，也有一说，骑马楼形似马鞍，故此得名。它坐北朝南，进深七檩，单檐歇山顶，楼上七间，楼下六间，东西各三间，下层中间有道路连通南北，

the Mid-Autumn Festival, this structure is an ideal site for moonlit fun.

To the east of the Yuxiu Building, separated by a fire lane, stands the Horseback Riding Building, which serves as the estate's guest lodgings. The term "Horseback Riding" signifies journeys and reflects the transient nature of life, highlighting the travels of visitors. Some believe the name was inspired by the building's saddle-shaped roof. This south-facing, two-story building features a design with seven purlins and a single-eave hip-and-gable roof. The upper floor has seven bays, while the lower floor has six, with three bays each on the eastern and western sides. A central passage connects the northern and southern sections of the structure. At the back, there is a three-sequential rear courtyard. The Horseback Riding Building is separated by a skywell from the adjacent Building 2 and 3 in the East Wing.

The layout reflects the traditional design typical of Yangzhou residences, featuring a "three-bay, two-wing, one-skywell" arrangement. The interconnected walkways form a guest suite complex that provides comfortable accommodations for visitors. This space was ideal for hosting guests; notably, the famous painter Huang Binhong (1865–1955) once stayed here. It is believed that the earlier Plum Building, which dates to the Twin Sophora Garden period, was located on this site before renovations made by the Ho family.

而楼后有院落三进，与其北之东二、东三两楼以天井相隔，为“三间两厢一天井”的扬州民居楼形式，又以楼道连接成组，足够待客留宿。当年，国画大师黄宾虹曾在此客居小住。据分析，双槐园时期的梅楼，旧址应该就在此处，何家入住何园时，对早期建筑进行了整修和再利用。

赏月楼
Moon-Viewing Building

牡丹厅“凤戏牡丹”砖雕
Brick carving "Phoenixes Frolicking Among Peonies, " on the roof of Peony Hall

Passing through the Horseback Riding Building to the north, one is greeted with the Grand Garden, the main northern garden of Ho Family Garden, which is the historical site for the Whistling Scholar's Mountain Retreat. Inspired by Su Shi's poem "The Mirage with a Preface," the garden revolves around the mythical theme of "fairy isles on the sea," making it the most imaginative part of He Weijian's design. This area is divided into the Eastern and West Gardens, featuring notable scenic spots such as the pond, the Wall-Backed Rockery, the Peony Hall (*Mudan Ting*), the Land Boat Hall, the Warter-Heart Pavilion, and the rockeries at the West Garden. Together, these elements create a poetic and

西园鸟瞰
Aerial view of West Garden

穿过骑马楼北行，映入眼帘的是何园的“大花园”，也是北部的主花园，此处历史上被称为“寄啸山庄”。这是一座以苏轼《海市（并叙）》为蓝本，以“海上仙山”为主题，颇具神话色彩的园林，也是何维键营造何园最具巧思的部分。寄啸山庄包含东、西两个部分，其间又有牡丹厅、船厅、贴壁假山、水池、水心亭、西园假山等重要景点，运用园林空间描绘了一幅海上行舟、访山求仙的浪漫图景。在东西花园之间的回廊上，嵌有《海市（并叙）》石刻，点明了寄啸山庄的意境所在。

东园鸟瞰
Aerial view of East Garden

romantic vision of a boat journey across the sea in search of immortal realms. The roofed walkway connecting the East and West Gardens is adorned with a stone inscription of the poem "The Mirage with a Preface," which beautifully captures the essence of this dreamlike landscape.

东园贴壁假山与近月亭

Wall-Backed Rockery of East Garden and Moon-Approach Pavilion

## 《海市（并叙）》

【宋】苏轼

予闻登州海市旧矣，父老云常出于春夏，今岁晚不复见矣。予到官五日而去，以不见为恨，祷于海神广德王之庙，明日见焉，乃作此诗。

东方云海空复空，群仙出没空明中。
荡摇浮世生万象，岂有贝阙藏珠宫。
心知所见皆幻影，敢以耳目烦神工。
岁寒水冷天地闭，为我起蛰鞭鱼龙。
重楼翠阜出霜晓，异事惊倒百岁翁。
人间所得容力取，世外无物谁为雄。
率然有请不我拒，信我人厄非天穷。
潮阳太守南迁归，喜见石廪堆祝融。
自言正直动山鬼，岂知造物哀龙钟。
信眉一笑岂易得，神之报汝亦已丰。
斜阳万里孤鸟没，但见碧海磨青铜。
新诗绮语亦安用，相与变灭随东风。

《海市（并叙）》
"The Mirage with a Preface"

## Scene 5: Wall-Backed Rockery

Upon entering the East Garden, visitors are immediately drawn to the Peony Hall. Originally built as the residence of the Ho family's steward, it was expanded after 1945. It features five bays with a hip-and-gable roof and is surrounded on three sides by verandas offering serene views of garden. True to its name, the hall is encircled by peony beds—both tree and herbaceous varieties—which create a vibrant floral scene in the spring. On its eastern wall, there is a striking brick relief, "Phoenixes Frolicking Among Peonies," known for its flowing lines and vivid detail.

Northeast of the hall lies a celebrated wall-backed rockery, coiling like a dragon along the garden wall. About 60 meters long, it includes a deep pond, winding stone paths, grottoes, and two viewing terraces: the Breeze-Greeting Pavilion and the Moon-Approach Pavilion. The rockery, resembling ink brushstrokes rising from the whitewashed wall, exemplifies Ming-Qing "stone-on-white" craftsmanship and is considered a masterpiece in Yangzhou.

Continuing northwest along the rockery, visitors arrive at the "Study Studio for Hanlin Scholar," named in honor of He Shenghao, who, after years of study there, was appointed to the Hanlin Academy during the Qing dynasty.

# 第 5 景　贴壁假山

进入东园，首先看到的是南部的牡丹厅，该建筑原为何氏管家的住房，抗日战争之后经改扩建而成。牡丹厅面阔五间，歇山顶，三面设置宽廊，厅周围有牡丹池，植以牡丹、芍药，花开时节一派姹紫嫣红的景象。东山墙上嵌砖雕“凤戏牡丹”图案，是扩建时由他处移建于此，砖雕线条流畅，层次丰富，具有较高的艺术价值。牡丹厅东北有假山，山体随园墙宛转盘曲，高下起伏约 60 米，下临深池，中有磴道和山洞，上有两座山亭，一名“接风”，一名“近月”，颇为风雅。这是最具扬州特色的贴壁假山，宛如嵌入墙体、依壁而生长的石山峰峦，白墙为纸，山石为墨，构成了立体的三维画卷。沿假山山道一直向西北，即可登上东园的“翰林公子读书楼”，与贯通全园的回廊联通汇合。当年园主何维键的长子何声灏在读书楼中苦读，终被光绪皇帝钦点翰林，故名。

贴壁假山鸟瞰
Aerial view of the Wall-Backed Rockery

## Scene 6: Land Boat Hall

North of Peony Hall and nestled between the Wall-Backed Rockery is the Land Boat Hall, the main building in the East Garden. Oriented toward the south, this hall historically served as the reception pavilion for gentlemen. The Land Boat Hall features a hip-and-gable roof with wide eaves and is three bays wide. It sits on a raised platform surrounded by short railings beneath its colonnade. The forecourt is paved with vertically set pebbles and tile fragments arranged in concentric ripples. The design exemplifies a garden technique called "dry garden simulating water" (*hanyuan shuizuo*), where landscaping creates the illusion of water features with the absence of actual water.

To the east of the Land Boat Hall, a winding waterway runs along the garden wall, creating the illusion that the hall is floating—echoing the "fairy isles on the sea" theme of the Whistling Scholar's Mountain Retreat. With windows on all sides, it offers panoramic views, enhancing the sensation of drifting through a a picturesque landscape. A couplet by late Qing official Li Zhongyu is displayed on the colonnade of the southern bay: "The moon as host, the plum as guests; / With flowers for walls, and a boat for rest." It reflects He Weijian's desire for retreat and

# 第6景　船厅

牡丹厅以北，比邻贴壁假山之处有一座花厅，被称作“船厅”，是东园的主体建筑。船厅坐北朝南，是当年的男宾花厅。该建筑宽廊歇山，面阔三间，下有平台，廊下短栏。厅前以竖瓦卵石镶铺地面，状若涟漪，此为造园艺术中“旱园水做”之法，即通过无水之园林营造出水波之意境。船厅东侧园墙下的水系，仿佛延伸到厅堂周围，将船厅托起，是寄啸山庄“海上仙山”意境的重要构成部分。船厅四面开窗，人在其中可以四周观景，仿佛船行山水间。南向明间廊柱上悬一副楹联，题为“月作主人梅作客，花为四壁船为家”，为光绪年间户部郎中李钟豫行书对联，与园主人回归林下、啸傲山水的造园初衷十分契合。园主何维键49岁退隐扬州，难免有壮志未酬之遗憾。后人又以“桴海轩”为名，典出《论语·公冶长》：“道不行，乘桴浮于海。从我者，其由与？”进一步强化船厅的立意。

船厅西侧有一条砖墙砌筑的通道，这是当年何园的北入口，来访宾客可以不经过南门，直接从北门进入花园。该入口门庭北向，是一座高大的磨砖门楼，大门之后为二门，即园子的正门，门楣上方嵌有“寄啸山庄”四字砖额，

his unfulfilled ambitions after his retirement at the age of 49. Later, the hall was renamed "Raft-on-Sea Veranda," referencing *The Analects*: "When the Way fails, I shall float on a raft over the sea…"—adding deeper symbolic meaning to the structure.

To the west of the Land Boat Hall lies a brick-walled passage—the original northern entrance of Ho Family Garden, allowing guests to enter directly from the north, bypassing the southern gate. This north-facing entrance features an imposing polished-brick gatehouse. Beyond the main gate stands a secondary gate, the garden's formal entrance, adorned with a carved brick plaque inscribed with the four characters of the Whistling Scholar's Mountain Retreat in clerical script, traditionally attributed to He Weijian.

Inside, the gate leads to a gallery system. Above it, an elevated gallery connects with the lower winding walkway, acting as a divider between the East and West Gardens. This structure features a wall embedded with an engraved poem titled "The Mirages with Preface" by the Song poet Su Shi. Through the openwork walls, latticed windows, and moon gates, guests can catch a glimpse of the West Garden's pavilions half-hidden among ancient rockeries and waterworks, evoking the ethereal motif of "fairy isles on the sea."

砖额隶书相传为园主人何维键所题。门内与穿廊相接，门上为串楼，与楼下廊扣合，即为分隔东西花园的复道回廊。廊壁间嵌苏轼撰书《海市（并叙）》，透过廊壁上的间隔花墙和空窗，可见西园内楼台掩映、假山苍古、水波浩渺，一派“海上仙山”的意境。

船厅
Land Boat Hall

## Scene 7: Water-Heart Pavilion

Passing through the multi-level corridor at the heart of the grand garden, one arrives at the West Garden—the centerpiece of the Whistling Scholar's Mountain Retreat's design. The West Garden is arranged around a central pond, with the principal structure situated along its northern shore. The western edge is embraced by artificial rockeries, and the southeastern sides are bordered by the multi-level walkways, forming a rectangular courtyard. At the pond's eastern quadrant stands the Water-Heart Pavilion (*Shuixin Ting*), also known as Little Square Isle (*Xiao Fanghu*), a square single-eave structure rising from a rectangular stone platform. A zigzag white stone bridge connects the pavilion to the southern bank, while a stepping-stone path made of Tai Lake limestone leads to the northern edge.

Its alternate name "Little Square Isle" is an allusion to "The Questions of Tang" in *The Book of Lieh-tzu*, which reads: "Beyond Bohai's eastern bounds lies a yawning chasm, / wherein rest five mystic isles—the third named Fanghu." A square pavilion poised in water embodies a "fairy isle on the sea" in mythology. Here, the aquatic pavilion extends the narrative voyage of the Land Boat Hall toward celestial realms.

# 第7景　水心亭

穿过大花园中部的复廊，来到西园，这里也是寄啸山庄的主景所在。西园以水池为中心，池北为主体建筑，池西为假山，东南两侧由复道回廊围合，形成长方形的院落。池中东偏水中，有水心亭，四方单檐，立于长方台基之上。亭西有三折白石平桥连接南岸，亭北有太湖石桥通于北岸。《列子·汤问》说：“渤海之东不知几亿万里，有大壑焉……其中有五山焉……三曰方壶”这种置于水中的方亭，在扬州园林中被称为“小方壶”，同样象征着“海中仙山”。这里用设立在水池之中的方亭表达花园部分从“船厅”驶向神山仙阁的意蕴。从园林空间营造的角度，建筑的作用变得非常突出，水心亭与桥的设置微微偏东，将水面分割为一大一小两部分，构成了疏密有致的层次。水心亭不仅是赏山水之景的最佳场所，较为饱满的体量也使其成为整个空间的中心。

水心亭冬景
Water-Heart Pavilion in winter

From the perspective of garden spatial layout, the architectural elements play a particularly prominent role. The Water-Heart Pavilion and its connecting bridges are deliberately positioned slightly east of center, dividing the water surface into a large and a small section, creating a harmonious composition of dense and sparse elements. The pavilion serves not only as the optimal vantage point for admiring the surrounding landscape, but its relatively substantial volume also establishes it as the visual and spatial focal point of the overall composition.

水心亭两侧的折桥与太湖石桥
The zigzag bridge and Tai Lake limestone bridge on the sides of Water-Heart Pavilion

# 第 8 景　蝴蝶厅

西园水池以北，有蝴蝶厅（又名“汇胜楼”）长楼倚北墙而筑、南向而立。建筑单檐歇山顶，面阔七间，中部三间稍向前突出，两侧间舒展如翼，状似蝴蝶，故名。七间长楼之东再接四间，建立梯道、走廊、半阁等，与复道回廊相接。正厅、两翼耳厅及半阁皆檐角飞举，建筑轮廓起伏变化，极富韵律。蝴蝶厅是西部主要厅堂，为园主收藏书画、会友宴客之所，因此又称汇胜楼宴厅。厅内东西板壁上刻有扬州八怪之一郑板桥的竹石图、江南才子唐寅花鸟画及刘墉等名家的书法。厅外廊柱罩槅上刻有八仙祝寿图。厅东侧与复道廊相连，廊壁设有漏窗，实现两侧景观的视觉互通与互动。可以想象当年宾客云集、杯盏交错、音律悠扬、轻歌曼舞，一番城市山林的好风光。汇胜楼西南侧有一座桂花厅，为 20 世纪 60 年代增建，因厅前种植金桂、银桂、丹桂、四季桂而得名。建筑坐西朝东，面阔三间，掩映于山石、朴树丛中，半隐半现，厅内悬康有为书“桂花飘香”横匾，每到金秋时节，满园芬芳。从室外磴道拾级而上，可从此转入曲折绕园的复道回廊。

# Scene 8: Butterfly Hall

To the north of the pond in the West Garden, the Butterfly Hall (*Hudie Ting*), also known as Splendor-Garnering Building (*Huisheng Lou*), stands against the northern wall, facing south. This elongated, two-story residential building features a hip-and-gable roof with a single eave and spans seven bays, with the central three projecting forward while the flanking wings extend gracefully outward, earning its name from its butterfly-like silhouette. To the east of the seven-bay hall, an additional four-bay structure, housing staircases, walkways, and a half-belvedere. The main hall, its flanking wings, and the half-pavilion all boast upturned eaves, enhancing the building's undulating, rhythmic roofline.

The Butterfly Hall is the great hall in the West Garden, serving as a venue for literati gatherings and banquets, as well as the owner's gallery for his collected paintings and calligraphic works, hence its alternate name Splendor-Garnering Building. Inside, its eastern and western wooden panels are engraved with notable artworks, including the bamboo-and-rock paintings by Zheng Banqiao (1693–1766, one of the Eight Eccentrics of Yangzhou), flower-and-bird paintings by Tang Yin (1470–1524, the legendary

蝴蝶厅
Butterfly Hall

Jiangnan polymath), and calligraphic works by Liu Yong (1719–1804) and other masters. The exterior walkway pillars and overhead beams feature intricate blessings of "Eight Immortals Celebrating Longevity." To the east, the hall connects seamlessly to the multi-level walkways, whose latticed windows allow glimpses of views on both sides. One can imagine the hall's heyday: guests admiring art, wind goblets clinking, melodies drifting through the air, and dancers moving gracefully—a microcosm of "forest in the city."

On the southwestern side is the Osmanthus Hall (*Guihua Ting*), an addition from the 1960s named after the fragrant osmanthus varieties planted in front. Facing east, this three-bay building is surrounded by rockeries and osmanthus groves. Inside the hall hangs a horizontal plaque with the words "Fragrance of Osmanthus," inscribed by Kang Youwei (1858–1927), a reformist scholar-official in the Late Qing dynasty. In autumn, the hall becomes a sensory highlight as the blossoms perfume the entire garden. Stone steps outside lead visitors upward to the multi-level walkway.

# 第 9 景　复道回廊

与蝴蝶厅隔水相对，可见一段复道回廊。所谓复道，是指双面回廊中设置的一道起分流作用的隔墙；而回廊就是扬州人俗称的“串楼”，乃串联各个院落和建筑之间的长廊。何园的复道回廊总长 1 500 多米，贯穿全园，有“有房必有廊，无廊不成房”之说。分隔东西花园和玉绣楼院落的复道回廊，是何园复道回廊中最为优美的片段，分上下两层，依墙而筑，壁间有什锦空窗上下排列，样式有梅花、海棠等，皆水磨

作为隔墙的复道回廊
Multi-Level Walkway, as a dividing wall

## Scene 9: Multi-Level Walkway

Opposite the Butterfly Hall lies a section of the so-called "multi-level walkway" (*Fudao Huilang*). The term "Fudao" denotes a double-layer walkway with a central dividing wall that helps direct the flow of movement, featuring both upper and lower levels. "Huilang" often called the "linking gallery" (*chuanlou*) in Yangzhou, is a

青砖边框，非常精美。西壁嵌有《颜鲁公三表》帖石，随读随行；复又折行于东廊之下，壁间嵌一部《唐人双钩十七帖》，有龙蛇飞动之势。回廊的设计增加了欣赏视角的变化，强化了西园的山池林谷之美。

高架空中的游径
Elevated walkways

long gallery that winds through the garden, connecting various architectural structures into a cohesive spatial sequence.

The multi-level walkway system at the Ho Family Garden stretches over 1,500 meters, echoing the saying "no structure stands without walkways; no walkway exists without a structure to serve." The section, which separates the Grand Garden and the courtyard surrounding the Yuxiu Building, was built along the garden walls. It features intricately designed polymorphic framing windows arranged in both vertical and horizontal patterns. These windows, framed with finely polished blue-gray bricks, showcase beautiful motifs, including plum blossoms and crabapple flowers, capturing the quintessence of the multi-level walkway system.

Its western wall is adorned with stone engravings of the "Three Memorials" by Yan Zhenqing (709–785), inviting visitors to appreciate the artistry of calligraphy when strolling along. As the walkway continues beneath the eastern gallery, the wall displays the calligraphic artwork titled "A Tang-Dynasty Rubbing of the Seventeen Model Letters in Double-Hook Style." Its impressive strokes flow with a dynamic, serpentine grace. The walkway layout enhances experiences by providing shifting views, enriching the layered landscape of the West Garden.

丰富的观景视角
The variety of viewing perspectives

西园假山
Rockeries at the West Garden

# 第10景　西园假山

蝴蝶厅西岸有湖石小山临水，山后有枫林、桂丛掩映入画。小山一路延伸向南，山势渐高，形成一座临水而起的高山，有磴道盘可上，俯瞰全院。假山上栽有两株白皮松，树龄已过百年，为光绪九年（1883年）寄啸山庄初建时所植，树干苍劲、姿态雄伟，在传统江南园林中十分罕见，增强了西园假山这座“海上仙山”的意趣。从园林设计的角度，此处将假山置于院落西南角，使池水南流处通道收窄，形成湖石水门。从池之北岸望去，池水犹有不尽之意。湖石假山的西面，又一座黄石假山拔地而起，与东西两处的湖石山子相接，构成了湖石山和黄石山交相辉映的奇特景观。根据史料推断，这段黄石假山建造的年代应早于湖石假山，大约在乾隆二十八年（1763年）前后，此时的园主为吴家龙之孙吴之黼，何氏在后续改造过程中保留了这段山体。

# Scene 10: Rockeries at the West Garden

On the western bank of the pond near the Butterfly Hall, there stands a Tai Lake limestone rockery backed by groves of maple and osmanthus, blending effortlessly into the landscape. As it extends southward, the rockery gradually rises, leading to a prominent hill with a winding stone path that invites visitors to ascend and take in views of the entire courtyard. Atop the rockery stand two Chinese white pines, planted in 1883 at the garden's inception. Their gnarled trunks and majestic forms are so impressive that they contribute to the vision of "fairy isles on the sea."

In the southwest corner of the courtyard, the rockery's placement serves both aesthetic and spatial functions. A narrowing water channel at the southern outflow forms a natural "limestone water gate," creating an illusion of endless water when viewed from the north. To the west of the Tai Lake limestone rockery stands a rugged Yellow Mountain rockery, forming a striking contrast with the limestone's porous elegance. Historical records indicate the Yellow Mountain rockery was built around 1763 by Wu Zhifu and later preserved by the Ho family for its artistic value.

行走至此，我们在何园里画出了一条接近 S 形的游线，从最具历史的部分——片石山房，一路行至北部花园，不仅领略了何园的变迁过程，也欣赏了何氏家族的不同生活场景。何园的游览还可以有很多组合，例如今天的游客大多数是从 1986 年新设的入口（徐凝门大街 66 号）进入，自北向南一路走入片石山房，可以看作是从当代走进历史。何园丰富的积淀和多样的景色赋予我们无穷无尽的行走、探索和体验之可能。

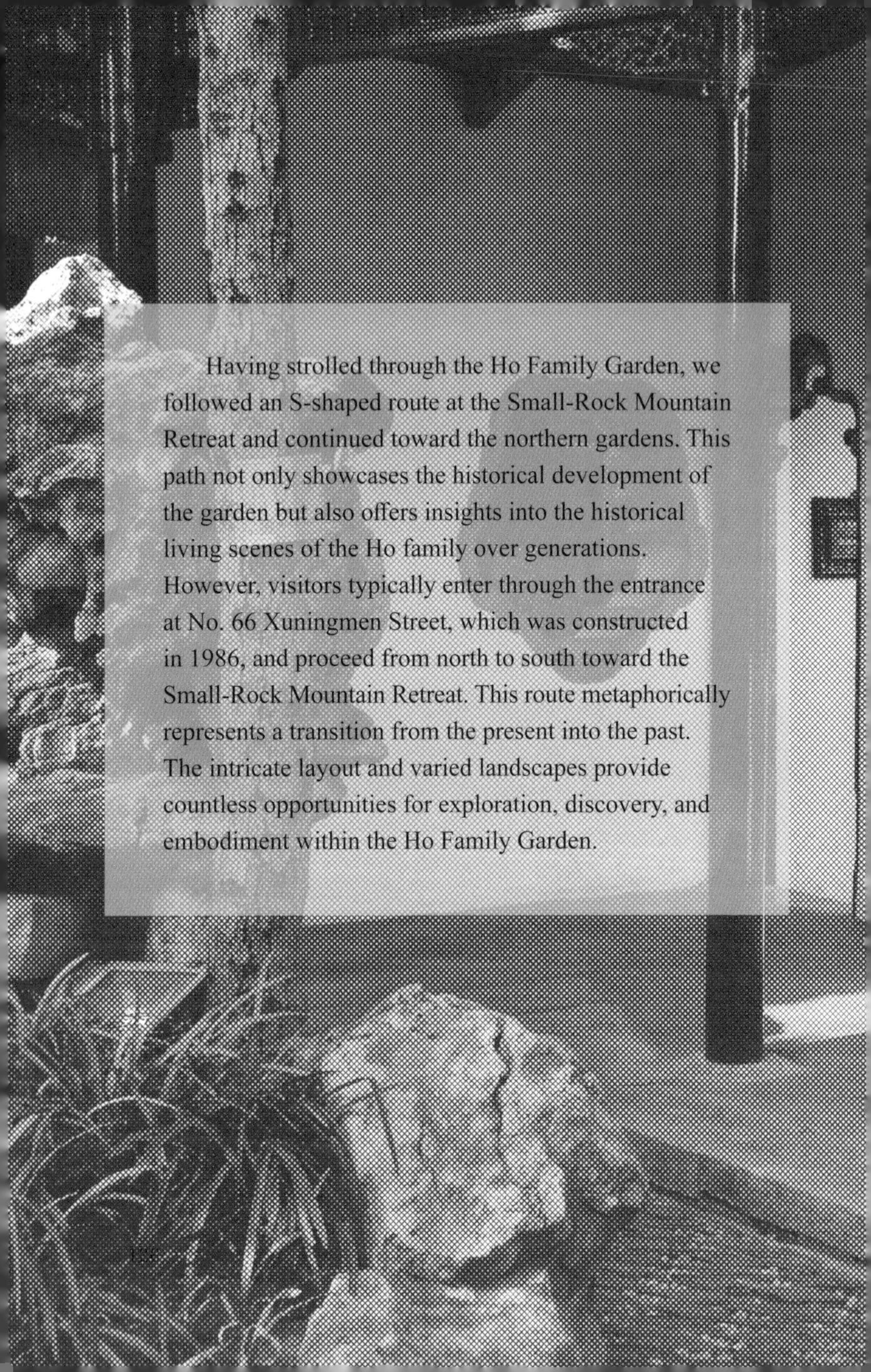

Having strolled through the Ho Family Garden, we followed an S-shaped route at the Small-Rock Mountain Retreat and continued toward the northern gardens. This path not only showcases the historical development of the garden but also offers insights into the historical living scenes of the Ho family over generations. However, visitors typically enter through the entrance at No. 66 Xuningmen Street, which was constructed in 1986, and proceed from north to south toward the Small-Rock Mountain Retreat. This route metaphorically represents a transition from the present into the past. The intricate layout and varied landscapes provide countless opportunities for exploration, discovery, and embodiment within the Ho Family Garden.

# 品园

APPRECIATIONS

## 回廊屈曲绕山楼

Linking Galleries Winding around Rockeries and Buildings

To fully appreciate the Ho Family Garden, one can stroll leisurely, discovering ever-changing views with each step; or take a moment to pause and savor the refined elegance embedded in its landscape. In the context of classical Chinese gardens, it shines like a radiant pearl, representing the concluding chapters in the development of classical garden art—a testament to both the enduring legacy and exquisite craftsmanship of this cultural tradition.

With a perfect balance between spaciousness and compactness, the garden boasts grand architecture, unique and weathered rock formations, and an ingeniously designed multilevel walkway system. The fusion of Chinese deco with exotic charm adds to its appeal. These distinctive features highlight the designer's exceptional creativity, making this garden a standout among the classical Chinese gardens.

However, the garden is more than a work of art; it is a living space crafted to accommodate daily life. While in the garden, imagine yourself as a member of the Ho family. Picture yourself practicing calligraphy and painting, burning incense in silent meditation, brewing tea, playing chess, feeding cranes, and listening to the soothing musical sounds of the *qin*. Enjoy pure talks with friends while sipping tea and wine. By immersing yourself as "a garden dweller," you'll gain a deeper understanding of its delicate landscape and craftsmanship, as well as the wisdom and spirit that this cherished home holds.

赏园之道，有动观与静观二法：游园宜动，缓步徜徉，步移景异，在空间转换中感受园林的节奏变化；品园宜静，驻足凝思，意与境合，在细节中体察造园者的匠心独运。若将何园置于中国古典园林的发展脉络中考察，可见其作为晚清园林艺术的代表，既继承了传统造园手法，又融入了时代特色，体现了古典园林成熟期的技艺与审美。

何园的布局遵循疏密对比的原则，建筑体量宏大却不失精巧，假山叠石兼具自然意趣与人工雕琢之美，复道回廊的设计巧妙连接不同空间，细部装饰则融合中西元素，形成独特的艺术风格。这些特征使何园在现存古典园林中具有鲜明的辨识度，也反映了造园者在功能与美学之间的平衡考量。然而，园林的本质并非纯粹的艺术品，而是服务于日常生活的空间载体。因此，在游览何园时，不妨从历史使用者的视角出发，想象其作为居住、社交、休闲场所的实际功能——或读书作画，或品茶对弈，或与友人雅集——从而更深入地理解园林设计与生活需求的紧密关联。这种基于实用性的空间营造，正是古典园林得以超越单纯观赏价值，成为文化载体的关键所在。

## Gardens Encircling Dwelling

The Ho Family Garden covers approximately 1.15 hectares, with half as gross floor area. A key challenge for its designer was how to reduce the feeling of high density and form a garden-like compound with an efficient overall layout. During the Ming-Qing dynasties, Yangzhou gardens mainly followed the traditional layout of residence in front (south), garden at the back (north, nicknamed "rear garden")—clearly separating movement and stillness, with layered spatial progression, as exemplified by the renowned Geyuan Garden (*Ge Yuan*). An alternative arrangement was the "side-garden layout," with a garden alongside the residence, such as the Lesser Winding Gully.

Yet, the Ho Family Garden charts an innovative design characterized by a wraparound layout—centering the residential quarters while enveloping them with gardens, creating a distinctive "gardens encircling dwelling" configuration. Thus, the Yuxiu Building, the main residence, is bordered by several gardens: the Grand Garden to the north, the Small-Rock Mountain Retreat to the southeast, the East Garden to the east, and the little garden of the Serenity Building to the west. The gardens and living spaces form interconnected yet self-contained systems, achieving a deliberate balance between private and public areas. This design caters to a harmonious blend of practical living and aesthetic enjoyment, historical continuity and innovative spatial

## 以园环居

何园的总面积约1.15公顷，其中建筑面积达6 000余平方米，建筑密度超过50%，这在江南私家园林中属于较高水平。面对如此高密度的建筑群，园主在规划时面临的核心挑战在于：如何通过空间组织手法化解建筑体量带来的压迫感，同时确保园林仍保持传统宅园的游赏品质。

从空间布局来看，何园突破了明清扬州园林常见的“前宅后园”（如个园）或“侧园式”（如小盘谷）的范式，创新性地采用了“以园环居”的环绕式布局——将住宅玉绣楼置于中心，四周以花园围合：东园以水景见长，片石山房以叠石取胜，北花园以建筑组群为特色，西侧怡萱楼小园则营造私密庭院空间。这种布局既保证了居住区域的独立性，又使每个功能区块都能便捷接入园林系统，形成“园中有宅、宅外有园”的嵌套结构。

从技术层面分析，何园通过三个策略实现了高密度下的舒适性：一是运用复道回廊串联各区域，在有限用地中拓展游线长度；二是通过假山、花窗等元素进行视觉分隔，降低建筑群体的实体压迫感；三是在细部处理上融合中西装饰语汇，以精致的艺术表现转移对建筑密度的注意力。这些手法

organization, as well as secluded privacy and expansive landscape immersion. Ultimately, it results in an enduring private garden that masterfully meets both functional needs and artistic aspirations.

Within the courtyards, space is often composed in a layered, centripetal manner. In some areas, such as the West Garden, a central scenic element is encircled by surrounding architecture; in others, like the East Garden, storied buildings form the core, embraced by rockeries, water, and flora. While rooted in classical spatial principles, the design reveals its distinctive character, subtly reinterpreting tradition with thoughtful innovation.

The West Garden excels in spatial composition, with a central water feature encircled by layered rockeries, pavilions, and flora. A two-story building to the north overlooks the garden, reinforcing a strong centripetal focus. Departing from the typical Jiangnan practice of placing a rockery directly opposite the main hall, the West Garden sets its rock formation to one side, creating a more spacious courtyard and a sense of boundless depth—an exceptionally rare layout among private Jiangnan gardens. The pond is edged with Lake Tai limestone, its contours softened by palms and magnolias that bridge water and architecture. A multi-level walkway surrounds the courtyard, resolving the constraints of high enclosing walls. Finished with whitewashed walls and polymorphic framing windows, the galleries reflect the water's shimmer, visually expand the space, and enrich the layering of the garden through framed vistas and borrowed views.

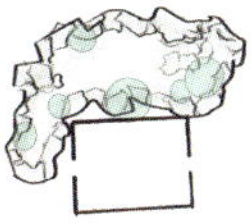

前宅后园
Front residence with rear garden

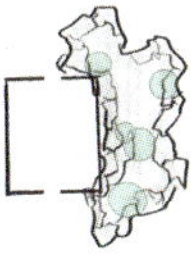

左宅右园
Left residence with right-side garden

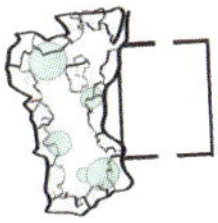

右宅左园
Right residence with left-side garden

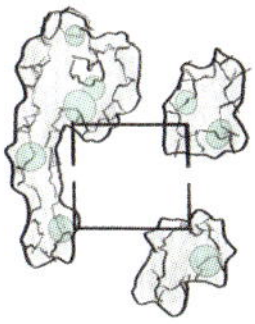

侧园式布局
Side garden layout

江南园林的院落布局
Courtyard layout styles of Jiangnan Gardens

“以园环居”的何园
Ho Family Garden—a residence surrounded by gardens

何园园林要素格局分析

Feature analysis of Ho Family Garden

玉绣楼内院

Inner courtyard of Yuxiu Building

共同构建了一个既满足居住需求，又具备高度艺术性的园林空间，体现了晚清园林在城市化背景下的适应性创新。

在何园的各个院落内部，也较多采用层叠向心式的布局，以主景为中心，周围环绕建筑，如西园；或以建筑为中心，周围环绕山水花木，如东园，而在尊重这种传统经典布局方式的同时，又有其独到之处。西园在空间处理方面最为精彩：中间布置较为集中的水面，环绕水面布置山石、建筑、花木，北侧筑两层楼房俯视全园，向心性和内聚感分外强烈。相较于其他园林，西园并未采用假山与主厅相对的传统"对景"布局，而是将体量较大的假山置于院落一侧，使得院落更加开敞，山体显露一角，增加望之不尽的感受，这在江南私家园林中是极为独特的手法。此外，水池以湖石驳岸收边，岸边山石掩映，点缀棕榈、玉兰，作为水池和建筑之间的过渡。院落周围复道回廊的设计巧妙地解决了高墙空间相对局促、压抑的难题，上下回廊皆为粉墙，局部嵌入什锦空窗，一方面能够反射水面并放大空间，另一方面通过空窗的透景与借景丰富了景观的层次。

## Multi-Story Buildings and Spacious Halls

The second defining characteristic of Ho Family Garden lies in its substantial architectural volume, predominantly featuring multi-story and belvedere-style structures. As noted by Chen Congzhou in *Gardens of Yangzhou* (1983): “The most distinctive architectural feature of Yangzhou gardens lies in their strategic use of elevated structures. The core building of the Whistling Scholar’s Mountain Retreat—the two-story, seven-bay Butterfly Building—integrated with its multi-level walkway system—directly overlooks the water, magnifying its monumental presence. This horizontally expansive architecture has juxtaposed against a vast aquatic surface, creating a striking visual paradigm.” The emergence of this distinctive architectural character has stemmed from a dual impetus: functional and spiritual imperatives, coupled with material and technological capacities.

Since the mid-Qing Dynasty, the Grand Canal waterscape in Yanghzhu has evolved into a central hub for social events and public leisure, leading to a layout reorganization of riverside gardens around the canal axis. This shift resulted in a distinctive architectural trend: multi-story pavilions were intentionally integrated into gardens to command panoramic views of the vibrant canal scenery. Notable examples include the Gaze-Upward Building (*Yangzhi Lou*) at the Slim Bamboo Garden (*Xiao Yuan*) and the Rivers-and-Mountains Vista Building (*Jiangshansiwang Lou*) at Refined

# 高楼敞厅

何园的第二个重要特征，就是建筑体量较大，且多为楼阁式建筑。陈从周先生在其《扬州园林》中提道："扬州园林在建筑方面最显著的特色，便是利用楼层，寄啸山庄的主体建筑，便是双层七开间的蝴蝶厅，与其两层复廊结合，直接面水，更显建筑之巨"。形成这一特征的原因，一方面是功能和精神需求，另一方面是建造的物质条件。自清代中叶以来，随着扬州运河水面空间成为社会事件与公共游览的中心，沿河园林开始以运河为核心组织其建筑空间布局，园林内部往往设立多层楼阁式建筑，以达到眺望运河风景的目的，如著名的筱园"仰止楼"、净香园"江山四望"楼等；扬州园林不再排斥大体量、多层化的建筑物，建造高大的主体建筑物反而成为一种趋势与风尚。

何园北部整体剖面图
Overall section of the northern part of Ho Family Garden

Fragrance Garden (*Jingxiang Yuan*). This era marked a paradigm shift in garden design in Yangzhou, as large-scale, multi-story structures became desirable and fashionable, rather than being seen with disdain.

In Yangzhou, many garden owners were wealthy merchants who often acquired official titles to enhance their social prestige. Although He Weijian and his son were reputed for their integrity, their appointments as Grain Circuit Intendant and Salt Transport Commissioner nevertheless brought them substantial personal wealth.

The prevailing philosophy underlying the design and construction of salt merchants' gardens was, most prominently, the pursuit of opulence—A means to flaunt wealth and cultivate an air of refined elegance. During the Kangxi-Qianlong era of the Qing dynasty, the salt merchants borrowed elements from imperial garden design to gain the favor of the emperor. This, in turn, helped them achieve their ambitions for official advancement and increased prosperity.

As a result, Yangzhou residential gardens would exhibit a consistent aesthetic grandeur in both architectural scale and material selection. They are characterized by spacious buildings and luxurious materials, and multi-level walkways, huge peaks and precious rocks are very common. A great hall often spanned seven bays, and there was a notable emphasis on architectural extravagance, with a preference for valuable *Phoebe zhennan* wood, and marble elements. These structures, while fulfilling the functional necessity of hosting numerous guests, crystallized the distinctive aesthetic of the Yangzhou gardens, setting them apart from their Suzhou counterparts.

此外，扬州园林的主人以富商为多，他们除拥有巨大的物质财富外，往往还会捐得一个官衔，以显耀其身份。何园的主人何维键与其子素以清正为官闻名，但在当时的社会背景下，“督粮道”和“盐运使”的职务为何氏家族带来了多元经济收入，积累了大量的财富。盐商园林设计与建造的主导思想中，最突出的特征便是追求豪奢，借以炫耀富有、标榜风雅。在清代康熙、乾隆年间，盐商们为了能得到皇帝的“御赏”以达到升官发财的目的，往往还会模拟一些皇家园林的手法。因此，扬州宅第园林在总体面貌上，建筑物的尺度、材料的品类，都追求高敞华丽，楼厅面阔有多至七间的，楼层复道、巨峰名石等要素更是十分常见。建筑用材以珍贵的楠木为佳，院落当中也不乏采用大理石的案例。这样的建筑在满足招待众多宾客的功能性需求之外，形成了扬州园林有别于苏州园林的独特风貌。

双层七开间的蝴蝶厅
Two-story, seven-bay Butterfly Hall

## Western Means for Chinese Ends

He Weijian, the owner of Ho Family Garden, once served in the Qing dynasty's diplomatic mission to France. Consequently, his garden design incorporated significant Western influences, blending European elements into both the residential layout and furnishings. This created a unique garden-residence spatial system where traditional Eastern philosophy intermingled with Western lifestyle concepts, endowing the garden with distinctive temporal characteristics.

The layout of the Ho Family Garden embraces variations and deviates from traditional norms. Rather than adhering to a central-axis layout, it responds to the site's topography by integrating remnants from earlier periods with new constructions. The Yuxiu Building, located along the western axis, stands out as the most representative example. It combines classical Chinese elements, such as gray-tiled roofs and brick walls, with European influences in spatial design and ornamentation.

The central courtyard of the complex is notably more expansive than the typical courtyards, allowing for better daylight access, improved ventilation, and more spacious activity zones. The two-story structures in the central courtyard consist of a series of interconnected rooms arranged into functional suites, with centrally located staircases and sliding-door partitions

# 洋为中用

何园主人何维键曾经担任过清王朝驻法国公使馆职务，因此，他的园林较多受到了西方文化影响，在宅居格局和家具设置上都加入了许多西洋的元素，构成了一个东方传统精神与西方生活观念交相杂糅的园居系统，赋予此园以鲜明的时代特征。何园在建筑布局上追求变化，不拘一格，没有采用传统中轴线式的前堂后屋布局，而是因地赋形，结合前代遗存进行保留、改造和新建，形成独特的园林面貌。最具代表性的片段是位于西轴线上的玉绣楼，其小青瓦屋顶、砖砌墙面都是传统的中式建筑风格，而空间格局和局部装饰则采用了许多欧式的做法。中部院落比一般的中式院落更宽敞，增加了采光量、通风性和活动空间，同时上下两层房间相互组合，形成多组套间，中间设置楼梯间，房间内以推拉门作隔断形成小套间，这种布局在当时是非常先进且实用的。此外，玉绣楼室内设置的壁炉、吊灯等元素都具有西式风格，不仅反映了时代特色，也在我国古典园林中开中西合璧风气之先河。

creating an adaptable subspace. This innovative and practical configuration was ahead of its time. In the Yuxiu Building, features like fireplaces and chandeliers have reflected Western influences, symbolizing a pioneering instance of Sino-Western architectural fusion in classical Chinese gardens.

具有西式风格的玉绣楼室内
Interior view of Yuxiu Building with western-style design

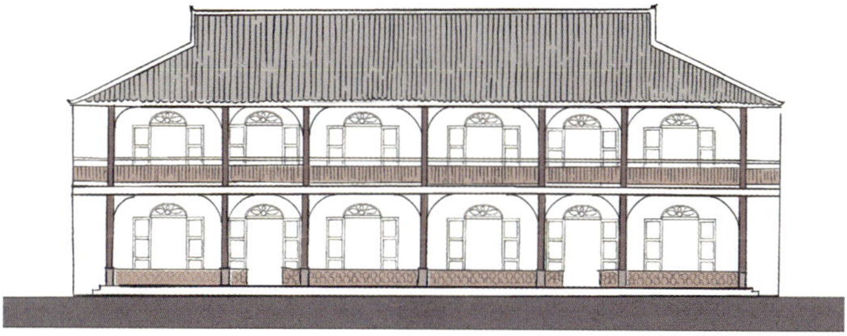

玉绣楼南部立面
South elevation of Yuxiu Building

## Wall-Backed Rockery Formation

The wall-backed rockeries, also known as "precipitous mountain" (*qiaobishan*) or "wall-crag" (*biyan*), feature a rock formation that resembles a cliff face against a wall. This represents an extraordinary technique in rock formation. Ji Cheng elucidated this method in *The Craft of Gardens*: "What is known as precipitous mountains are built up against walls so that the whitewashed surface acts as the painting paper and the rocks as the painting upon it. The designer should follow the natural cracks in the rock, imitating the brushwork of the great predecessors. The rocks should be planted with ancient plum trees, fine bamboo, as well as the pines from Huangshan. Looking at this scene through a round window is akin to seeing a view reflected in a mirror."

Yangzhou is situated on flat plains and lacks native rock resources suitable for garden landscaping. As a result, stones are sourced from other regions, often transported as ballast in salt boats. This dependence on external sources has led to the use of smaller yet valuable materials. The limitations of local resources have inspired innovative approaches to rock formation, where small stones are carefully arranged along the walls of gardens and structures based on their shape, color, grain, texture, and intrinsic character to create a harmonious composition. To reinforce and stabilize these structures, strips of stone and iron braces are embedded within the design. Once they are mortared and sealed,

## 贴墙作山

贴墙作山，即“峭壁山”，或称“壁岩”，是一种非常独特的堆叠园林假山的技法。计成在《园冶》中说道：“峭壁山者，靠壁理也。藉以粉墙为纸，以石为绘也。理者相石皴纹，仿古人笔意，植黄山松柏、古梅、美竹，收之圆窗，宛然镜游也。”扬州地处平原，当地没有可供堆叠假山的石料，石材皆取自外地，大多利用盐船压舱运回，因此石料一般较小，且非常珍贵。这一条件的限制反而成就了扬州园林壁岩假山的鲜明特色。壁岩即借助园林围墙或建筑墙面的支撑，用小石拼接，根据石形、石色、石纹、石理、石性等组合成整体，中以条石、铁器支挑，加固嵌填后浑然成章,形成完整的假山形态,核心技巧被称为“包镶之法”。壁岩假山的石块之间一般采用传统胶合材料固定，或石灰加细砂和糯米汁，凝结后呈现微红色，多用于黄石堆叠的假山；或在其中加入草灰，则颜色显灰白，适用于太湖石假山。嵌缝则采用阴嵌之法，即见缝不见灰，既能显出其壁石凹凸多态，也能模仿自然裂纹，形成浑然一体的效果。壁岩假山手法自然逼真，节省用材，并能够充分利用园林空间。

何园即是此类叠山的典型代表，其东园的贴壁假山突出展示了扬派叠石的技巧，假山依墙而筑，一路攀行，状若游龙，将原本非常局限的高墙深院，转化成古朴自然的山水环境。在有限的

the rockery presents a seamless and cohesive appearance, a method known as "material-encapsulation technique" (*baoxiang zhifa*).

The rocks used in rockery construction are typically bonded together with various materials. For instance, a mixture of lime, fine sand, and glutinous rice paste, which sets to a faint reddish hue, is commonly used for the Yellow Mountain rockeries. Another option is to use plant ash, which produces a pale gray tone and is applied to Lake Tai limestones. The joints are carefully filled using the "recessed mortaring technique" (*yinqianfa*), which conceals mortar and minimizes its visibility. This approach highlights the natural contours of the stonework while artfully simulating geological fissures, resulting in a harmonious and organic unity. This wall-crag rockery technique exemplifies authentic mimicry, material efficiency, and spatial optimization.

The rock formation at the Ho Family Garden serves as a prime example of these methods and techniques, most notably demonstrated in those of the East Garden, which exemplifies the distinctive Yangzhou style. Built against the walls, the rockery ascends along a winding path, transforming the once-enclosed courtyard into a rustic, natural landscape. Despite spatial constraints, the garden skillfully incorporates layered horizontal and vertical dimensions, featuring peaks, caves, stone paths, pavilions, and lush vegetation. This layered design evokes a dynamic and poetic representation of mountainous terrain.

The western summit of the rockery at the Small-Rock Mountain Retreat employs intricate stone-piecing techniques, embodying

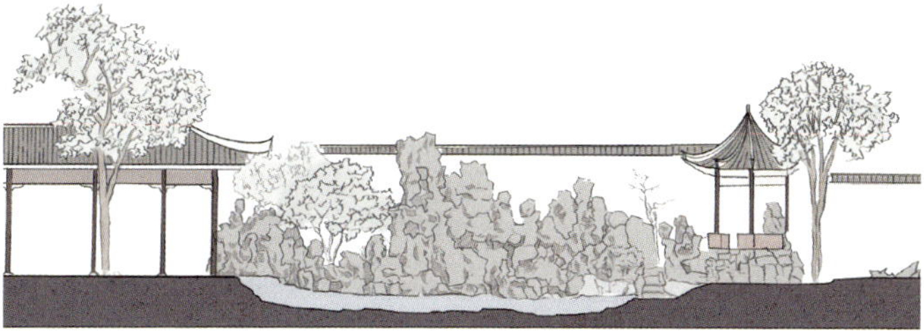

东园贴壁假山立面图（东段）
Elevation of the east section of the Wall-Backed Rockery in the East Garden

东园贴壁假山：扬派叠石代表作
The Wall-Backed Rockery in the East Garden exemplifies the distinctive Yangzhou style

the painting principle articulated by Shi Tao: “Peaks and texture-strokes unite—The strokes born from the peaks’ essence.” This composition unfolds like an archaic landscape handscroll rendered in three dimensions. This rockery, featuring its impressive peaks and deep caves, appears entirely natural. The consistent textures of the stones, the balance between solid forms and open spaces, as well as the compact layout, all reveal a remarkable level of artistic skill and ingenuity. In May 2013, a full-scale replica was constructed and is now on display at the Chinese Garden Museum in Beijing.

高墙深院，因山化境

Rockery ascends, walls become nature

山体空间中又穿插了水平向和垂直向的层次，峰峦、洞室、磴道、凉亭、植物等要素无一不有，形成重叠递进的山势变化，极尽诗情画意。片石假山西侧主峰运用小石拼镶，体现了石涛“峰与皴合，皴自峰生”的画理，宛如一幅古朴的山水长卷。片石山房虽然峰高洞深，但浑然天成，完全看不出人工的痕迹，其石纹的统一、体量的虚实对比、布局的紧凑，需要极高的艺术修养和巧妙的技法构思方可实现。2013 年 5 月，片石山房作为中国园林“教科书”，被原样复制并由北京中国园林博物馆收藏。

## Multi-Level Walkway System

The wall-backed rockery epitomizes the Yangzhou's distinctive artistry in rockwork formation, while the multi-level walkway system in the Ho Family Garden stands as unparalleled innovation—A singular masterpiece absent not only from other Yangzhou gardens but all classical Chinese landscapes. The interlocking corridors weave through the entire compound with a labyrinthine complexity, incorporating multi-level pathways—both upper and lower—as a three-dimensional circulation. Its vertical circulation and roofed structure allow for climate-proof transit between courtyards and ensure all-weather access. Furthermore, the double-way in certain sections effectively manages visitor flow through strategically placed diversion nodes. This interconnected pathway system has been celebrated as "the proto-form of Chinese interchange bridges."

Moreover, the system serves as an essential landscape interface with the landscape and a viewing platform, skillfully integrating straight galleries, curved walkways, encircling passages, and stacked corridors. In the pool area at the East Garden, where buildings envelop the water on three sides, the distinctive view walkways offer elevated scenes, enhancing spatial depth and alleviating the oppressiveness of tall walls through the layered layout. Visitors can enjoy a dynamic sequence as the landscape changes across different elevations.

## 回廊复道

如果说贴壁假山是扬州叠山特色及艺术水平的反映，那么复道回廊则是何园园主的独创作品，这不仅是扬州园林中的孤例，在国内其他古典园林中也未见此手法。何园的复道回廊曲折回环，四通八达，上下交叠，贯穿全园。这一独特构筑首先体现的是其实用功能的优势，它为宅第园林创造了一个便捷、安全和舒适的全天候立体交通空间，无论风霜雨雪，都可以畅行于各个院落，复道还能够有效起到分流作用，把园林四通八达之妙发挥得淋漓尽致，是中国园林中绝无仅有的精彩景观，被誉为“中国立交桥的雏形”。同时，复道回廊提供了重要的景观界面和赏景空间，它包括直廊、曲廊、回廊、叠廊等多种样式的巧妙组合，西园水池三面环楼，回廊的设置既可从其上俯视园林，又为院落增加了景观的层次，减弱了高墙带来的压抑感。廊的东南两面墙上开有什锦洞窗和水磨漏窗，绕廊赏景，景随步换，在高低不同的视点中，园景产生不同变化，营造出俯瞰园景的高视点连续观赏途径，形成不同一般的赏景序列。廊壁间利用磨砖窗洞达到互相“借景”的效果，显得空灵深远，这是中国园林利用分隔与透景扩大空间感受的典型手法之一。何园将中国传统园林中的廊元素进行了充分运用和大胆创新，游览何园，如果不在楼上楼下曲曲折折、高高低低地走上一回，是很难领略它的独特韵味的。

The southeast openwork walls are adorned with polished lattice and polymorphic framing or perforated windows, creating the effect of shifting scenes with each step. This design facilitates a sequential revelation, forming a continuous high-viewpoint pathway that embodies a unique spatial narrative in Chinese gardening.

The walkway walls employ polished-brick lattice windows to achieve mutual "borrowed scenery" effects, imparting an ethereal sense of depth—a quintessential Chinese garden technique that amplifies spatial perception through strategic partitioning and framed vistas. The Ho Family Garden elevates and innovates upon the traditional Chinese garden walkway with vertical dynamism (multi-level winding passages) and spatial drama (ascent and descent rhythm). To truly grasp its genius and experience its poetic essence, one must traverse every undulating level, where each turn stages a new dialogue between architecture and nature.

复道回廊提供的景观界面
A viewing platform

# Openwork Walls

Openwork walls (*Huaqiang*) refer to perforated garden walls adorned with latticed or framing windows—A signature Jiangnan gardening technique that enables mutual scenery borrowing and spatial permeability. Ji Cheng, in *The Craft of Gardens*, notes that "Within gardens, walls bordering flower beds, water edges, pathways, or encircling rockeries should employ polished-brick lattice walls. These serve dual purposes: shielding interiors from outside view while maintaining refined aesthetics, achieving timeless elegance that delights the senses."

Compared to other regions in Jiangnan, the window designs found in the openwork walls in Yangzhou gardens display unique characteristics in both typology and craftsmanship. These windows can be categorized into two types: perforated windows (*dongchuang*, solid openings) and latticed windows (*louchuang*, geometric brick tracery). The prominent motifs include shapes such as book scrolls, quatrefoils, full moons, and fans. Notably, the pine-crane design, common in Suzhou, is absent here. The technical skill involved in creating these windows is evident in the polished-brick masonry and the modular assembly of large geometric units. This craftsmanship is characterized by precise grinding, innovative structural designs, and complex compositions, showcasing remarkable artistic achievements.

# 透空花墙

花墙是对园林中各个院落墙体上进行透空处理，塑造形态各异的花窗，从而形成相互借景和渗透，是江南园林的常用手法。《园冶》中说道："园内花前、水畔、路旁、环山的围墙，宜用水磨砖作漏砖墙，可以避外隐内，雅致合时，令人欣赏"。与江南其他地区相比，扬州园林中的花墙特点鲜明，以类型来言，有洞窗和漏窗两种形式，形状有书卷、亚方、圆月、扇面等，不见或少见松鹤图案的花窗。同时，扬州的花窗多以水磨砖叠砌成大块几何图形，其磨制之精、叠砌之巧、构图之难、样式之美，都达到了极高的艺术水准。

水心亭与花墙
Water-Heart Pavilion and Openwork Walls

The openwork wall at the Ho Family Garden is one of the distinctive features of Yangzhou gardens. These perforated partitions, such as the east wall of the Peony Hall and the divider between the Eastern and West Gardens, showcase refined craftsmanship. The latter is constructed with precisely fitted polished blue-gray bricks and features an array of delicate lattice patterns. Its central section boasts two picture-frame windows adorned with crabapple-blossom borders made from dressed square bricks, while small gates punctuate the northern and southern ends of the wall. This ingenious design not only obscures the pavilions, rockeries, and flora of the West Garden to create layered vistas but also offers glimpses of the scenery through apertures. It evokes the poetic allure of "a veiled lute player," enhancing visitors' curiosity through controlled visibility. This concept embodies the principle of "concealed elegance" that characterizes the landscaping of Jiangnan gardens.

In addition, the Ho Family Garden features a remarkable series of "Polymorphic Framing Windows" (*shijin kongchuang*). A notable example can be found in the multi-level walkway to the south of the West Garden's pool, where two tiers of these windows display intricate designs, including floral motifs such as plum blossoms, crabapple, and water caltrop, as well as cultural symbols like Ruyi scepters, lotus flowers, and sword hilts. Each polished square-brick window functions as a unique painting, with a complete variation in patterns—no two windows are identical.

何园的透空花墙是扬州园林最为出色的代表作之一，园内用于分割庭院空间的墙体，如牡丹厅东面的隔墙，东、西花园之间的隔墙等，都砌作透空花墙。就东、西花园的隔墙而言，以水磨小青砖拼砌成不同的透空花窗样式，十分精细雅致。此段花墙的中部有两面空窗，采用水磨大方砖拼砌作海棠花式边框，墙之南北两端开设小门，这一设计既能将山庄主景的西部园中楼台山池花树遮藏起来，增加园景的层次，又隐约透出园中景色，形成犹抱琵琶半遮面的含蓄美，能够充分激发游人一探佳境的雅兴。此外，何园中还有一类“什锦空窗”，例如西园内水池之南的复道廊墙上，有上下两排空窗，以水磨大方砖拼砌成梅花、海棠、菱花、如意、莲花、剑环等样式，极为精美，每扇空窗都能自成一景，面面不同，宛若画图。

什锦空窗
Polymorphic Framing Windows

More than just a silent witness to history, Ho Family Garden is a living vessel of cultural continuity. Sunset spills over its blue-gray bricks and ebony tiles, as this century-old masterpiece whispers enduring tales. The 2024 exhibition, “Wandering Through Jiangnan Gardens,” held in Versailles, carried its allure beyond borders, letting Ming aesthetics converse with Western formalism. As a verse says: “Though visitors depart, the towers remain vibrantly alive; / The ancient shines anew in a modern light.” Here, where East meets West in the New Era, the garden still breathes—its rockeries pulse with life, and its waters mirror infinite skies.

Its true charm lies not only in architectural splendor or landscape elegance, but in layered history, humanistic depth, artisan brilliance, and Eastern philosophy. To wander its paths in ink is but a fleeting delight—stepping forward, you may discover your own Ho Family Garden.

“Wandering Through Jiangnan Gardens” Exhibition in Versailles, France

随着夕阳的余晖洒在何园的青砖黛瓦上，这座百年名园仿佛在低声诉说着它的故事。何园，不仅是历史的见证者，更是文化的传承者。2024 年的“行走江南园林”法国站展览，让何园的魅力跨越了国界，与西方园林展开了一场跨越时空的对话。正如书中所言，“人去楼未空，旧园焕新容”，何园在新时代的舞台上，依然焕发着勃勃生机。

何园的魅力不仅在于其建筑的华丽和景色的优美，更在于其丰富的历史故事、深厚的人文内涵、精湛的工匠智慧和深刻的东方哲思。纸上悠游终觉浅，让我们迈开步子，去寻找自己的何家花园！

“行走江南园林”法国站展览现场照片

# 附录 1

## 何园古树名木统计表

| 编号<br>No. | 树名<br>Common Name | 学名<br>Botanical Name | 科属<br>Family, Genus | 位置<br>Location | 树龄（截至20<br>Tree Age (up t |
|---|---|---|---|---|---|
| 1 | 白皮松<br>Chinese White Pine | *Pinus Bungeana* | 松科<br>松属<br>Pinaceae<br>Pinus | 赏月楼北面假山半山腰处（东西两侧）<br>Mid-slope of the Rockery (East & West Flanks), north of the Moon-Viewing Building | 154 |
| 2 | 瓜子黄杨<br>Willow-leaf Boxwood | *Buxus Microphylla var. Sinica* | 黄杨科<br>黄杨属<br>Buxaceae<br>Buxus | 桂花厅东南侧花台<br>Flower terrace southeast of the Osmanthus Hall | 154 |
| 3 | 瓜子黄杨<br>Willow-leaf Boxwood | *Buxus Microphylla var. Sinica* | 黄杨科<br>黄杨属<br>Buxaceae<br>Buxus | 船厅西南侧花台<br>Flower terrace southwest of the Land Boat Hall | 154 |
| 4 | 罗汉松<br>Buddhist Pine | *Podocarpus macrophyllus* | 松科<br>罗汉松属<br>Podocarpaceae<br>Podocarpus | 片石山房假山东侧<br>East slope, rockery of Small-Rock Mountain Retreat | 349 |

# APPENDIX I

## Catalogue of Ancient and Notable Trees at the Ho Family Garden

| 树名<br>Common Name | 学名<br>Botanical Name | 科属<br>Family, Genus | 位置<br>Location | 树龄（截至2025年）<br>Tree Age (up to 2025) |
|---|---|---|---|---|
| 木绣球<br>Chinese Snowball Viburnum | *Viburnum macrocephalum* | 忍冬科<br>荚蒾属<br>Podocarpaceae<br>Podocarpus | 片石山房假山东侧<br>East slope, rockery of Small-Rock Mountain Retreat | 154 |
| 广玉兰<br>Southern Magnolia | *Magnolia grandiflora* | 木兰科<br>木兰属<br>Magnoliaceae<br>Magnolia | 玉绣楼中心庭院内<br>Central courtyard, Yuxiu Building | 154 |
| 紫薇<br>Crape Myrtle | *Lagerstroemia indica* | 千屈菜科<br>紫薇属<br>Lythraceae<br>Lagerstroemia | 赏月楼南侧假山上<br>Rockery, south of the Moon-Viewing Building | 154 |
| 女贞<br>Glossy Privet | *Ligustrum lucidum* | 木樨科<br>女贞属<br>Oleaceae<br>Ligustrum | 赏月楼南侧<br>South of the Moon-Viewing Building | 154 |

# 附录 1

## 何园古树名木统计表

| 编号<br>No. | 树名<br>Common Name | 学名<br>Botanical Name | 科属<br>Family, Genus | 位置<br>Location | 树龄(截至2(<br>Tree Age (up t |
|---|---|---|---|---|---|
| 9 | 朴树<br>Chinese Hackberry | *Celtis sinensis* | 榆科<br>朴属<br>Cannabaceae<br>Celtis | 赏月楼西侧<br>West side of the Moon-Viewing Building | 154 |
| 10 | 桂花<br>*Osmanthus* | *Osmanthus fragrans* | 木樨科<br>木樨属<br>Oleaceae<br>Osmanthus | 赏月楼南侧<br>South of the Moon-Viewing Building | 154 |
| 11 | 石楠<br>Chinese Photinia | *Photinia serrulate* | 蔷薇科<br>石楠属<br>Rosaceae<br>Photinia | 赏月楼北侧<br>North of the Moon-Viewing Building | 154 |
| 12 | 女贞<br>Glossy Privet | *Ligustrum lucidum* | 木樨科<br>女贞属<br>Oleaceae<br>Ligustrum | 桂花厅南侧假山上<br>Rockery, south of the Osmanthus Hall | 154 |

# APPENDIX I

## Catalogue of Ancient and Notable Trees at the Ho Family Garden

| 树名<br>Common Name | 学名<br>Botanical Name | 科属<br>Family, Genus | 位置<br>Location | 树龄（截至2025年）<br>Tree Age (up to 2025) |
|---|---|---|---|---|
| 瓜子黄杨<br>Willow-leaf Boxwood | *Buxus Microphylla var. Sinica* | 黄杨科<br>黄杨属<br>Buxaceae<br>Buxus | 东大门北侧牡丹花台上<br>Peony Terrace, north of the East Gate | 154 |
| 桂花<br>*Osmanthus* | *Osmanthus fragrans* | 木樨科<br>木樨属<br>Oleaceae<br>Osmanthus | 船厅西南侧（东西两侧）<br>Southwest of the Land Boat Hall (East & West Flanks) | 144 |
| 桂花<br>*Osmanthus* | *Osmanthus fragrans* | 木樨科<br>木樨属<br>Oleaceae<br>Osmanthus | 赏月楼北侧桂花群（东部5群）<br>Osmanthus Grove, north of the Moon-Viewing Building (East Section 1-5) | 144 |
| 丝棉木<br>Maack's Spindle Tree | *Euonymus maackii* | 卫矛科<br>卫矛属<br>Celastraceae<br>Euonymus | 东大门<br>East Gate | 104 |

# 附录 2
## 何园大事记

| 年份<br>Year | | 事记<br>Event |
|---|---|---|
| 1703 | 康熙四十二年 | 吴家龙初辟片石山房(又名“双槐园”)，园中叠石相传<br>清初名画僧石涛营构。<br>Wu Jialong first established the Small-Rock Mountain Retreat, also known as Twin Sophora Garden. The rockeri in the garden were traditionally attributed to Shi Tao (164 1707), a renowned painter-monk from the Qing dynasty. |
| 1736<br>—<br>1795 | 乾隆年间 | 当时的园主人修葺此园，于园北增筑石山，由湖上莲性<br>(今法海寺)僧牧山和尚设计并布置。<br>The owner renovated the garden and constructed a rocker on the northern side. The design and construction were entrusted to Monk Mushan from the Lianxing Temple by t lake, now known as Fahai Temple. |
| 1818 | 嘉庆二十三年 | 熊之垣《花间笑语》载，吴之黼为片石山房主，园景荒败<br>只存听雨轩、水榭，时称双槐茶园。<br>According to Xiong Zhiyuan's book *Jests Among Flowers* Wu Zhiyun owned the Small-Rock Mountain Retreat. The garden fell into disrepair and was referred to as the Twin Sophora Tea House. The only remaining structures were t Rain-Listening Veranda and the Waterside Gazebo. |

# APPENDIX II

## Chronicle of the Ho Family Garden

| 年份<br>Year | | 事记<br>Event |
|---|---|---|
| 838 | 道光十八年 | 钱泳《履园丛话》载，片石山房“为一媒婆所得，以开面馆，兼为卖戏之所，改造大厅房，仿佛京师前门外戏院式样，俗不可耐矣”。<br>According to *A Collection of Accounts at the Lü Garden* by the Qing literatus Qian Yong (1759–1844), the Small-Rock Mountain Retreat was purchased by a matchmaker and transformed into a noodle shop and a makeshift theater. The great hall was remodeled in the style of the low-quality theaters found outside the Front Gate in Beijing. |
| 883 | 光绪九年 | 湖北汉黄德道道台、江汉关监督何维健在扬州花园巷购片石山房（“双槐园”旧址），并在其西及北修建“寄啸山庄”，山水楼堂环抱园居院落，为江南园林之孤例。<br>He Weijian, the Circuit Intendant of the Han-Huang-De area in Hubei and the Superintendent of Hankou Customs, purchased the Small-Rock Mountain Retreat, located at the former site of the Twin Sophora Tea House on Garden Lane in Yangzhou. He constructed the Whistling Scholar's Mountain Retreat to the west and north of the original structure. The garden features a harmonious blend of gardens, rocks, waterscapes, pavilions, and halls surrounding the residential courtyard, making it a unique example among the gardens in Jiangnan. |
| 886 | 光绪十二年 | 著名画家黄宾虹做客何园。<br>The renowned painter Huang Binhong (1865–1955) visited the garden. |

# 附录 2
## 何园大事记

| 年份<br>Year | 事记<br>Event |
| --- | --- |
| 1935 | 何维键长子何声灏率子孙搬回扬州何园居住。<br>He Shenghao, the eldest son of He Weijian, returned with his family to reside in the garden. |
| 1937 | 日寇全面侵华，何声灏以 74 岁高龄率全家祖孙三辈二十余人离开扬州何园逃难，经汉口、桂林、广州，再绕道香港，于 1938 年冬抵上海寓居法租界中。<br>With the outbreak of Japanese full-scale agression in Chi 74-year-old He Shenghao led three generations of his fam over twenty members, on a harrowing exodus from their ancestral home, Ho Family Garden. Their difficult journe took them through Hankou, Guilin, and Guangzhou, befor detouring via Hong Kong and reaching safety in the Frenc Concession of Shanghai in the winter of 1938. |
| 1944 | 何氏后人将何园整体出售给殷汝耕。<br>The garden was sold to Yin Rugeng, a notorious collabora with the Japanese invaders. |
| 1945 | 抗战胜利后，国民政府以没收敌伪财产为由接收何园，在此兴办“祝同中学”及“淮安中学”。<br>Upon the victory in the War of Resistance against Japanes Aggression, the Nationalist Government claimed the gard under the pretext of confiscating the enemy and puppet regime properties and repurposed the classical estate to house the newly established Zhutong High School and Huai’an High School. |

# APPENDIX II

## Chronicle of the Ho Family Garden

| 年份<br>Year | 事记<br>Event |
| --- | --- |
| 1949 | 何园先为苏北军区所在地，后为江苏省军区文化速成中学（二〇速中）、六机部七院第十研究所（七一〇所）和六机部七院第二十三研究所（七二三所）使用。<br>The garden served as the headquarters for the Northern Jiangsu Military District Command. Later, it housed three institutional users: first, the Accelerated Cultural Middle School of Jiangsu Provincial Military District, known as the "20th Accelerated School." This was followed by Research Institutes 10 and 23 of the 7th Academy of the Sixth Ministry of Machine-Building (i.e., No. 710 & No. 723 Research Institutes of China Ship Research and Development Academy). |
| 1959 年 10 月<br>October, 1959 | 扬州市人民政府将何园园林部分交扬州市园林所整修并对外开放。<br>The landscape section was transferred to the Yangzhou Municipal Landscape Administration for conservation and public access. |
| 1961—1962 | 同济大学建筑系青年教师陈从周率同济大学建筑系三十余名师生在扬州进行古民居建筑考察，其间在何园住宅部分所在地花园巷发现石涛和尚叠石遗作——片石山房。<br>Chen Congzhou, a young faculty member of the Architecture Department at Tongji University, led a team of more than thirty teachers and students to survey the historic dwellings of Yangzhou. During their fieldwork, they uncovered a remarkable rockery masterpiece by Shi Tao, known as the Small-Rock Mountain Retreat, located on Huayuan Lane adjacent to the residential courtyard. |

# 附录 2
## 何园大事记

| 年份<br>Year | 事记<br>Event |
| --- | --- |
| 1985 年 3 月<br>March, 1985 | 江苏省人民政府公布何园为省级文物保护单位<br>The garden was designated a Provincial-Level Cultural Heritage Site by the Jiangsu Provincial Government. |
| 1985 年 9 月<br>September, 1985 | 在国务院关心下，何园住宅部分及“片石山房”由中国舶工业总公司七二三研究所(七二三所)移交给扬州市林局进行整修，“片石山房”修复后，于 1989 年 10 对外开放。<br>Under the auspices of the State Council, both the residential section and the Small-Rock Mountain Retrea were transferred from the Research Institute 723 of Chi State Shipbuilding Corporation to the Yangzhou Munici Landscape Administration for comprehensive restoratio The garden reopened to the public in October 1989. |
| 1988 年 1 月<br>January, 1988 | 国务院公布何园为全国重点文物保护单位。<br>The garden was designated one of the Major Historical Cultural Sites Protected at the National Key Level. |
| 1999 | 9 月 14 日，中国科学院院士、何氏后人何祚庥首次到何参观。<br>On September 14, He Zuoxiu, a descendant of the Ho family and an academician at the Chinese Academy of Sciences, visited the garden for the first time. |

# APPENDIX II

## Chronicle of the Ho Family Garden

| 年份<br>Year | 事记<br>Event |
| --- | --- |
| 2002—2003 | 先后修缮了玉绣楼等建筑，并对外开放。<br>The Yuxiu Building and other historic structures were renovated and opened to visitors. |
| 2005 | 11 月 8 日，何园管理处召开何家海外赤子故里寻根问祖叙情会。同年，扬州何园更名为扬州市何园管理处，何园被全国旅游景区质量等级委员会评为国家 AAAA 旅游景区。<br>On November 8, the Administration Office of Ho Family Garden hosted a homecoming gathering for overseas descendants of the Ho family, helping them trace their ancestral roots and reconnect with their cultural heritage. In the same year, the garden was officially designated as the Yangzhou Administration Office of Ho Family Garden and recognized as a National AAAA-Level Tourist Attraction by the National Committee for Tourist Attraction Quality Rating. |
| 2012 | 12 月 16 日—18 日，扬州文物专家到何园开展文物普查。<br>From December 16 to 18, the cultural heritage experts in Yangzhou conducted a systematic inventory of historical artifacts at the garden. |
| 2020 | 10 月 25 日，《数字化园林遗产图录：扬州何园》新书首发式在何园蝴蝶厅举行。这是何园首次开展数字化遗产保护的工作成果，由同济大学景观学系杨晨副教授主持。<br>On October 25, the new book launch for *Garden Heritage, Digital Document: Heyuan Garden / Yangzhou* was held at the Butterfly Hall of Heyuan. This marked the first digital surveying and mapping project of Ho Family Garden, led by Prof. Yang Chen from Tongji University. |

# 图片来源
# Figure Source

| 图片 / 图纸页码<br>Pages of Figures / Drawings | 来源<br>Source |
| --- | --- |
| 第 18、23 页<br>Pages 18, 23 | 周仰“不朽的林泉”系列摄影作品<br>From “Faërie” series by ZHOU Yang |
| 第 24-25、39、53、60、79、107、110-111、117、121、150-151、165、172-173 页<br>Pages 24-25,39,53,60,79,107,110-111,117,121,150-151,165,172-173 | 扬州市何园管理处 提供<br>Provided by the Administration Office of Ho Family Garden |
| 第 28 页，拉页内（下）<br>Page 28, inside of the foldout page (bottom) | 杨于莺 绘<br>Illustrated by YANG Yuying |
| 第 27、33、37、43、54、62-63、66-67、71、82-83、85、87、89、93、94-98、102-103、112、115、118-119、125、141-142、147、157、161、163、167 页<br>Pages 27,33,37,43,54,62-63,66-67,71,82-83,85,87,89,93,94-98,102-103,112,115,118-119,125,141-142,147,157,161,163,167 | 杨晨 摄 / 提供<br>Photo/provided by YANG Chen |
| 第 49、73、122、155、156-157、161 页<br>Pages 49,73,122,155,156-157,161 | 王毓麟 摄<br>Photo by WANG Yulin |
| 第 58-59 页，拉页内（上）<br>Pages 58-59, inside of the foldout page (above) | 杨午燕 绘<br>Illustrated by YANG Wuyan |
| 第 141、145、151、155 页<br>Pages 141,145,151,155 | 刘俊豪 绘<br>Illustrated by LIU Junhao |
| 第 166 页<br>Page 166 | 聂阳阳 提供<br>Provided by NIE Yangyang |

# 编后记

“城市行走”已然成为了当下探索城市的一种生活方式：年轻人在行走中发现自我，中年人在行走中释放自我，老年人在行走中重塑自我。

“城市行走”跨越国界，它是人与世界对话的一种方式。无论是在巴黎的街巷、伦敦的河畔，还是江南的园林，人们都能通过行走感知历史、触碰文化，用脚步丈量世界，用心灵连接彼此。

“城市行走”跨越时光，它是人与历史对话的一种方式。从古代城墙到现代街区，从百年园林到摩登建筑，行走让我们在时间的长河中，追溯过往、感知当下，赓续历史文脉，谱写未来华章。

“行走江南园林”系列丛书的诞生，源于我们对“城市行走”这一独特文化形式的别样思考和实践。在策划初期，我们希望通过实体的行走来探访那些尚且存世的名园佳苑，更希望通过精神的“行走”来深入理解园林所承载的江南文化与生活智慧。这既是对“城市行走”理念的延续，也体现了高校出版社在当下的责任与担当　　用专业的视角、系统的内容、独到的设计，为读者提供一个重新认识城市与历史的平台。

机缘巧合的是，这一项目在萌芽之时，正逢《奔流：从上海出发——全球城市人文对话》首季法国场系列活动的组织。借此契机，概念得以成为现实，“行走江南园林”巡展以江南文化和园林美学为中华文化国际传播的重要抓手，成功把江南园林带到了凡尔赛，用“东园西鉴”的方式来体现文明交流互鉴，得到了广泛认可。在2024年中法建交60周年之际，我们踏出了新的一步：通过实体模型、艺术作品、数字投影、昆曲表演、学术讲座等多样的形式，用有限的展览空间创造出游园的无尽体验——这些早已突破了出版的边界，也促使我们重新思考了出版的意义。

“内容为王”向来是出版的核心竞争力和价值所在，在当下知识传播的媒介形式发生质变之时，内容的表达形式也不应该只拘泥于图书。但出版人作为“知识生产的组织者”的这个角色定位，不会因为时代的改变而改变；孵化出多元形式的优质内容来回馈社会，是我们通过“行走江南园林”项目想要实现的创新——“始于出版，不止于出版”。与此同时，我们也希望继续传承和发扬同济大学出版社“城市行走书系”的初心——图书超越记录之用，予人启发，提供平台，是一种分享城市、生活与文化的独特方式。

感谢所有为这套丛书出谋划策、付出努力的大伙伴和小伙伴们，特别鸣谢本套丛书的主编章明教授的启发、指导和鼓励，同时，感谢同济大学宣传部、建筑与城市规划学院以及中华外译与国际传播研究基地的支持，感谢基地主任陈琳教授团队的英文翻译。最后，感谢扬州市何园管理处为本项目的展览活动和图书编撰提供的支持与帮助。孔子言："志于道，据于德，依于仁，游于艺。"我们以"书"为媒，以"行"会友，以"园"寄意，岂不妙哉？

呈现在各位面前的"行走江南园林"不仅仅是一套书籍，更是一场邀约。我们希望能与大家一起，在说园、游园和品园的过程中，找到属于自己的"桃花源"。

"行走江南园林"项目统筹组

2025 年 5 月

# Epilogue

“City Walk” has become a modern way to explore the city: the young discover themselves through walking, the middle-aged find relief in strolling, and the elderly rediscover themselves through meandering.

“City Walk” transcends borders—an open dialogue between humanity and the world. Whether meandering through boulevards and alleys in Paris, strolling along riverbanks in London, or tracing the winding paths of Jiangnan gardens, one perceives history through footsteps and touches culture with every stride. This is how we measure the earth’s contours with our soles and bridge souls across civilizations.

“City Walk” transcends time—a trans-temporal dialogue between humanity and history itself. From ancient city walls to modern blocks, from centuries-old gardens to contemporary architecture, walking allows us to navigate the river of time: tracing echoes of the past, awakening to the pulse of the present, and weaving unbroken cultural threads that illuminate future narratives.

The forthcoming series stems from our practice of “City Walk” as a means of cultural immersion. During the initial planning stages, we sought not only to physically explore the surviving masterpieces of classical gardens but also to embark on a spiritual journey, delving deep into the Jiangnan culture and life philosophy embedded within these landscapes. This endeavor is both an extension of the “City Walk” concept and a reflection of a university press's mission in the contemporary era: to offer readers a platform for rediscovering cities and history through intellectual rigor, systematic content, and innovative design.

By a remarkable coincidence, this project began to take shape precisely as the first season of "Our Water" (Shanghai × Paris) project was being organized—a timely convergence that proved pivotal. Seizing this opportunity, we held the touring exhibition "Wandering Through Jiangnan Gardens," which successfully introduced classical Chinese gardens to Versailles and served as a means of communicating Chinese culture internationally. When East meets West in these gardens, the dialogue among civilizations becomes readily accessible, leading to widespread acclaim for this exhibition. In 2024, marking the 60th anniversary of Sino-French diplomatic relations, we embarked on a groundbreaking endeavor: transforming finite exhibition spaces into boundless garden experiences through a multisensory tapestry of physical models, artistic installations, digital projections, Kunqu opera performances, and academic dialogues. This innovative fusion has not only transcended conventional publishing boundaries but has fundamentally redefined the very essence of cultural dissemination.

"Content is King" has always been the cornerstone of publishing—its core competitiveness and enduring value. Yet, in an era where the methods of knowledge dissemination have significantly evolved, the expression of content should no longer be limited to books alone. Despite these changes, the vital role of publishers as "curators of knowledge production" remains timeless. This project seeks to innovate by incubating high-quality content in diverse forms—"starting with publishing, but not confined to it." At the same time, we remain committed to upholding the original vision of the *CityWalk Book Series*, as envisioned by Tongji University Press, where books go beyond mere documentation to inspire, create platforms, and serve as a unique medium for sharing insights about cities, life, and culture.

Our deepest gratitude goes to all the colleagues and partners who shared knowledge and efforts in creating this series. We would like to express our special thanks to Professor Zhang Ming, editor-in-chief of the series, for his inspiration, guidance, and unwavering encouragement. We are also profoundly grateful for the continuous assistance from various departments at Tongji University, including the Publicity Department, College of Architecture and Urban Planning, as well as the Center for Translating China for Global Communication. Finally, we would likc to extend our gratitude to Professor Chen Lin's team for their translation of the English version. We also appreciate the invaluable support provided by the Administration Office of the Ho Family Garden, Yangzhou. A Confucian adage goes that, "Devote yourself to the Way, depend on virtue, rely on benevolence, and wander in the arts." Inspired by this spirit, we have chosen books as our medium, walking as our fellowship, and gardens as our muse—could there be any pursuit more sublime?

The Project Coordination Team for<br>"Wandering Through Jiangnan Gardens" Project<br>May 2025

图书在版编目（CIP）数据

行走江南园林．何园篇：汉英对照 / 章明主编；杨晨著．-- 上海：同济大学出版社，2025. 7. -- ISBN 978-7-5765-1702-6

Ⅰ．TU986.625

中国国家版本馆 CIP 数据核字第 2025ZC3600 号

上海文化发展基金会图书出版专项基金资助项目

# 行走江南园林 · 何园篇

# Wandering Through Jiangnan Gardens: Ho Family Garden

杨晨 著 YANG Chen

出 品 人　金英伟
项目统筹　顾旭峰 高晓辉
项目宣传　聂阳阳 蒋卓文
英文翻译　陈琳
策划编辑　袁佳麟
责任编辑　袁佳麟 王毓麟
责任校对　徐逢乔
插画设计　杨午燕 杨于莺
装帧设计　苏诗淼 付雨彤
出版发行　同济大学出版社 http://press.tongji.edu.cn/
（地址：上海市四平路 1239 号 邮编：200092 电话：021-65985622）
经　　销　全国各地新华书店
印　　刷　上海安枫印务有限公司
开　　本　787 mm × 1092 mm　1/32
印　　张　6
字　　数　108 000
印　　次　2025 年 7 月 第 1 版
版　　次　2025 年 7 月 第 1 次印刷
书　　号　ISBN 978-7-5765-1702-6
定　　价　65.00 元

仅以此书纪念我在扬州的故友谢青桐，感谢他在我探索前行道路上的分享、鼓励与启迪。

*This book is dedicated to the memory of my dear friend Xie Qingtong from Yangzhou—Whose shared wisdom, encouragement, and inspiration illuminated every step of my journey.*